Conduction in Non-Crystalline Materials

Conduction in Non-Crystalline Materials

SIR NEVILL MOTT

Emeritus Professor of Physics,
Cavendish Laboratory, University of Cambridge

SECOND EDITION

CLARENDON PRESS · OXFORD
1993

Oxford University Press, Walton Street, Oxford OX2 6DP
Oxford New York Toronto
Delhi Bombay Calcutta Madras Karachi
Kuala Lumpur Singapore Hong Kong Tokyo
Nairobi Dar es Salaam Cape Town
Melbourne Auckland Madrid
and associated companies in
Berlin Ibadan

Oxford is a trade mark of Oxford University Press

Published in the United States
by Oxford University Press Inc., New York

First edition 1987
Second edition 1993

A catalogue record for this book is available from the British Library

Library of Congress Cataloging in Publication Data
Mott, N. F. (Nevill Francis), Sir, 1905–
Conduction in non-crystalline materials / Nevill Mott.—2nd ed.
1. Energy-band theory of solids. 2. Free electron theory of metals.
3. Amorphous substances—Electric properties.
4. Semiconductors. I. Title.
QC176.8.E4M66 1993   530.4'16—dc20   92-32384
ISBN 0–19–853979–7

Typeset by Integral Typesetting, Great Yarmouth, Norfolk
Printed and bound in Great Britain by
Bookcraft (Bath) Ltd, Midsomer Norton, Avon

Preface to second edition

This second edition differs from the first in several ways. It contains a much fuller account of a condensed electron gas, both in the region of 'weak localization' where the work of Bergmann and of Altshuler and Aronov are reviewed and the behaviour of an electron gas near the metal–insulator transition. Here I follow my work with Kaveh in assuming that the equation deduced for weak localization can be extrapolated to the (Anderson) metal–insulator transition, an assumption partly justified by its success in reproducing the observations.

The question of a minimum metallic conductivity is treated more fully than in the first edition. It followed from the scaling theory of 1979 that such a quantity does not in general exist, but there are exceptions; at the lowest temperatures it can be observed in some doped semiconductors when the metal–insulator transition is induced by a strong magnetic field, and the quantity is relevant for liquids, where quantum interference appears to disappear.

In view of the book by Kamimura and Aoki (1989), which covers particularly the properties of an electron gas in the localized, non-conducting regime, we have not included much discussion of these problems here.

Finally, we have added a short chapter on the copper oxide super-conductors. Except for $Y_1B_2Cu_3O_{7-\delta}$ with $\delta = 0$, the carriers in most of these materials move in a non-period field, and if, as the author believes, the carriers above and below the critical temperature are bosons, we need to consider Anderson localization for them.

Cambridge   N.M.
November 1992

Preface to first edition

There are now available several books about non-crystalline solids. E. A. David and I published *Electronic processes in non-crystalline materials* in 1971, with a second edition in 1979. S. R. Elliott's *Physics of amorphous materials* appeared in 1984 and Richard Zallen's *Physics of amorphous solids* in 1983. There are also books in the Springer series in which several authors share, for instance *Fundamental physics of amorphous semiconductors* (ed. F. Yonezawa, 1980), and *The physics of amorphous silicon* (ed. J. O. Joanopoulos and G. Lucovsky, 1984). On the related problem of highly doped semiconductors there is the excellent book of Shklovskii and Efros (1984).

My aims in adding yet another book mainly on the theory are the following. The first is to provide a reasonably small book which I hope can serve as an introduction both for students of experiment and of theory. The next is to bring my earlier book with Davis up to date; so much has happened since 1979 that much of the theory presented there is in need of revision. And finally, I hope to show more fully than in my earlier books that non-crystalline semiconductors are by no means the only materials to which the concepts described here can be applied; vitreous silicon dioxide, amorphous metals, and impurity bands in doped semiconductors will play an equal role.

I have called this book *Conduction in non-crystalline materials* because it is here that the theory differs most from that for crystals. But conduction includes photoconduction, effects of a magnetic field, and so on, and so optical and magnetic effects can by no means be excluded.

Finally I would like to thank several colleagues who have looked through all or parts of the manuscript, particularly E. A. Davis, M. Kaveh, and M. Pepper.

Cambridge   N.M.
April 1986

Contents

1 Introduction

1.1. Conduction in crystalline systems

Before the appearance of quantum mechanics we had little understanding of why some solids, such as the metals, were good conductors of electricity and others were not. The Hall effect gave a measure of the number of free electrons in a metal, the Hall constant R_H being, according to theories then available, equal to $1/nec$, n being the number of electrons per unit volume, e the electronic charge, and c the speed of light. From the experimental values of this quantity it appeared that the number of free electrons in a metal was of the same order as the number of atoms. In insulators, on the other hand, all electrons seemed to be stuck; none were free to move. This could not be explained, nor could many other properties of solids. A major success of electron theory was, however, the explanation of the Wiedermann–Franz ratio of the electrical (σ) to the thermal (K) conductivity of metals (Lorentz 1905); theory gave

$$K/\sigma = 2(k_B/e)^2 T$$

where k_B is the Boltzmann constant and T the absolute temperature. This is in fair agreement with experiment. But outstanding problems were, why the mean free path, particularly at low temperatures, is so large in comparison with the interatomic distance, and why the free electrons do not contribute a large term ($\frac{3}{2}nk_B$) to the specific heat, in addition to that ($3Nk_B$) from the thermal vibrations. Here n is the number of electrons per unit volume and N the number of atoms.

Pauli in 1926 first applied the Fermi–Dirac statistics to account for the analogous problem of the paramagnetism (why the tree electrons do not contribute a large paramagnetism equal to $n\mu^2/k_B T$, μ being the magnetic moment of the electron). Arnold Sommerfeld, who for decades had presided over the outstanding school of Munich, saw Pauli's paper in proof (Hoddeson and Baym 1980) and extended it to the problem of the specific heat. If Fermi–Dirac rather than classical Boltzmann statistics describe the energies of the electrons, these will be spread over a range of energies equal to E_F, the Fermi energy, of the order of several electron volts (eV) and therefore large compared with $k_B T$; only a fraction $\sim k_B T/E_F$ of them would have any thermal energy. Thus the internal energy would be $\sim n(k_B T)^2/E_F$ and the

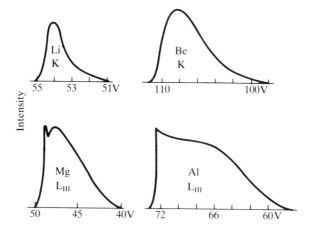

Fig. 1.1. X-ray emission bands of light metals (O'Bryan and Skinner 1934).

specific heat $\sim k_B^2 T/E_F$. That the electron energies were indeed spread over a range of several eV was first shown experimentally through the X-ray emission band of light metals by O'Bryan and Skinner (1934); some of their results are shown in Fig. 1.1. A small specific heat linear in T at low temperatures was first observed in silver in 1934 by Keesom and Kok in the Netherlands; their results agreed well with the theory for a free-electron gas (Fig. 1.2).

Sommerfeld, however, did not address the problem of an electron moving in the periodic field of crystal. This was first done in a seminal paper by

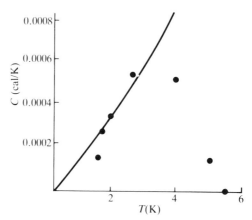

Fig. 1.2. Linear term in the specific heat of silver (Keesom and Kok 1934). The full line shows the theoretical value.

Fig. 1.3. Showing a typical wave function of type (1.2) using a tight binding model; the diagram shows the real or imaginary part. (From Ashcroft and Mermin 1976, p. 185, Fig. 10.7.)

Bloch (1928), who considered the Schrödinger equation

$$\nabla^2 \psi + \frac{2m}{\hbar^2}(E - V)\psi = 0 \tag{1.1}$$

where $V (= V(x, y, z))$ is a periodic function of position. Here ψ is the wave function and $\nabla^2 = \partial^2/\partial x^2 + \partial^2/\partial y^2 + \partial^2/\partial z^2$. The solutions of such an equation are

$$\psi = \exp(i\mathbf{k} \cdot \mathbf{r})u(x, y, z) \tag{1.2}$$

where $u(x, y, z)$ has the same periodicity as V. A typical solution is shown in Fig. 1.3. This solution represents a plane wave modulated by the crystal field. It does not show any scattering; the wave vector $\mathbf{k}$, and thus the momentum of an electron, has a constant value. Bloch could therefore argue that the long mean free paths observed at low temperatures were to be expected. Resistivity is a result of deviations from the perfect crystal lattice resulting from thermal vibrations or from the presence of impurities. As regards thermal vibrations, scattering should be proportional to the square of the displacement of an atom from its mean position, so that above the Debye temperature the resistivity will be proportional to $k_B T/f$, where fX is the restoring force on an atom for a displacement X. Bloch also showed how the possible energies $E(k_x, k_y, k_z)$ of an electron were divided into zones (the Brillouin zones) with gaps between them.

A. H. Wilson (1931) first pointed out that the model gives a clear distinction between metals and insulators. In metals one or more zones are partly occupied, so that a surface in k-space (the 'Fermi surface') separates occupied from empty states; the Fermi surface is a sphere only if $E(\mathbf{k})$ is a function only of the magnitude of $\mathbf{k}$, for instance equal to const $|\mathbf{k}|^2$. In insulators all zones are full or empty, and there is a gap between the highest occupied band (the valence band) and the lowest empty band (the conduction band). Thus electrons are not 'stuck'; on the contrary they are mobile, but unless an electron is removed from the valence band thermally or by the absorption of radiation, exactly as many electrons move in one direction as in the other. Moreover, an (extrinsic) semiconductor is a material in which impurities (dopants) provide occupied states with energies just below the conduction band, so that at room temperature many or most of the electrons are free to move.

As is well known, this model of Bloch and Wilson has survived for the treatment of semiconductors throughout the enormously important developments of germanium and silicon technology, which have taken place since the end of the Second World War. We shall show, however, particularly in Chapter 7, that, since the existence of zones depends on the assumption that the material is crystalline; it is not adequate to account for the (obvious) property of oxide glasses, that they are transparent, and therefore that a gap exists. A gap, then, cannot depend essentially on the properties of the solution of eqn (1.1) with a crystalline form of $V(x, y, z)$. Gaps must exist for some forms of $V(x, y, z)$ appropriate to non-crystalline materials. This will be apparent when we consider Anderson's random potential introduced in Chapter 3.

At this point we emphasize a difference between the theories of semiconductors and of metals. In the former, we have to do with a low concentration of electrons in the conduction band; interaction between them is rarely important, except in the formation of 'excitons' when an electron and positive hole form a bound pair. On the other hand, in metals the interaction term e^2/r_{12} is large, r_{12} being the distance between pairs of electrons. Early work neglected this, but it was essential to show that, in spite of this large term, a sharp Fermi energy and a sharp Fermi surface existed. Jones, Mott, and Skinner were the first to show the former, in an attempt to explain the sharp upper limit to the X-ray emission bands shown in Fig. 1.1, while the discussion of the Fermi surface is due to Landau (1957). The sharp upper limit is thought to exist in amorphous as well as crystalline metals, though the Fermi surface does not. Interaction between electrons leads to surprisingly small effects in the electrical properties of crystalline metals, perhaps the most important being a small term in the resistivity proportion to T^2 resulting from electron–electron collisions (Landau and Pomeranchuk 1936; Baber 1937). In non-crystalline metals this interaction is more important, as we shall see in Chapter 5.

We finish this section by setting down a few formulae which will be used later in this book.

In metals, the conductivity depends only on the properties of electrons at the Fermi surface (or in a non-crystalline material at the Fermi energy). For a spherical Fermi surface, one can write for the conductivity σ

$$\sigma = ne^2\tau/m,$$

where n is the number of electrons per unit volume, e, the electronic charge, τ the time of relaxation, and m the effective mass. The time of relaxation is related to the mean free path l by the equation

$$l = v\tau$$

where v is the velocity of an electron at the Fermi surface. In terms of l, we can write

$$\sigma = ne^2 l/mv$$

$$= ne^2 l/k_F \hbar$$

where k_F is the wave vector at the Fermi surface. Since each state in k-space is associated with a volume $8\pi^3$, for n we have

$$\tfrac{1}{2} n = (4\pi/3)k_F^3/8\pi^3,$$

where the factor $\tfrac{1}{2}$ comes from the two spin directions; then

$$\sigma = 4\pi k_F^2 e^2 l/12\pi^3 \hbar.$$

$4\pi k_F^2$ is the Fermi surface area S_F, so

$$\sigma = e^2 S_F l/12\pi^3 \hbar, \tag{1.4}$$

an equation which will be used in this book.

Naturally these equations depend on the assumption that, if we write $a = n^{-1/3}$ so that a is the mean distance between the electrons, l is large compared with a. This is called the weak-scattering limit. One of the main themes of this book is the problem of what happens when $l \sim a$. A principle due to Ioffe and Regel (1960) suggests that it cannot be smaller; we shall show why in § 3.2.

1.2. Non-crystalline systems

In this book we treat conduction in the following systems.

1. *Impurity conduction in doped and compensated semiconductors.* In silicon or germanium lightly doped with, say, phosphorus, the energy, denoted by ε_1, needed to remove an electron from the phosphorus into the conduction band is of order

$$\varepsilon_1 = me^4/2\hbar^2 \kappa^2.$$

Here, m is the effective mass and κ the dielectric constant. The wave function of an electron attached to the phosphorus P^+ ion is like that of a hydrogen atom with radius a_H given by

$$a_H = \hbar^2 \kappa/me^2.$$

For concentrations n of dopant greater than n_c where

$$n_c^{1/3} a_H \simeq 0.25,$$

the material behaves like a metal, in the sense that the conductivity tends to a finite value as the temperature tends to zero. This kind of metal–non-metal transition cannot be treated without considering interaction (Chapter 4). But for smaller concentrations, conduction can take place by direct transfer of electrons from one centre to another if the material is *compensated*. This means that it contains a smaller concentration of an acceptor (for instance boron). Then all the acceptors will be negatively charged; some of the donors will be neutral and contain an electron and others positively charged and thus empty. The form of conduction which results is called 'impurity conduction'. It has been extensively studied (cf. Shklovskii and Efros 1984). Since the radius a_H is in general large compared with the lattice parameter, distortion of the lattice by a trapped electron and such phenomena as polaron formation and Stokes shift have in general a negligible influence. These systems, investigated at the very lowest available temperatures, are therefore ideal for seeking to understand in its simplest form the motion of electrons in a non-periodic field. But this is a problem where electron–electron interactions play a major role. We devote Chapters 4 and 5 to this phenomenon.

2. *Non-crystalline metals.* These often show interesting electrical properties, including a negative temperature coefficient of resistance. They can be described through concepts of the same kind as those developed for heavily doped semiconductors, and are treated in § 2.4 and Chapter 5.

3. *Polaron motion.* This is a phenomenon of importance in both crystals and some non-crystalline materials, and is discussed in Chapter 6.

4. *Non-crystalline semiconductors.* The earliest investigations were on the properties of amorphous selenium, used by the Xerox company in the electrostatic copying process known as xerography, and those of the late Boris Kolomiets and co-workers in St Petersburg on the chalcogenide glasses (see Kolomiets 1964). These are glasses with, for example, a composition such as As_2Te_3 and also alloy glasses containing arsenic, tellurium, silicon, and germanium. More recently amorphous silicon containing upwards of 5 per cent of hydrogen has been extensively studied. These materials have—in the main—the same coordination number as in the crystal, if this exists; in this they are unlike liquid and amorphous metals, for which there is no integral coordination number. Thus As will normally be bounded to three neighbours, Si or Ge to four, and Te and Se to two. The explanation of their properties is based on the assumption that conduction and valence bands exist, as in a crystalline material, but that the lowest states in these bands can act as traps; they are said to be localized. Also a fully coordinated material would have a gap in the energy spectrum between the bands, but 'defects'—that is points where the coordination is abnormal—do give rise to (localized) states in the gap. A major achievement of Kolomiet's school was to show that the chalcogenide glasses cannot be doped; the conductivity depends little on purity. On the other hand doping is possible in deposited

films of hydrogenated silicon, and has led to the development of p–n junctions in amorphous silicon and their use as photocells. These non-crystalline semiconductors are treated in Chapter 7.

5. *Liquid metals.* These have properties which differ somewhat from those of amorphous metals. The classical Ziman theory is described in Chapter 2, and effects resulting from short mean free paths in Chapter 8.

6. *Vitreous SiO$_2$ and its formation by oxidation of silicon.* Vitreous silicon dioxide has one of the largest band gaps known (~ 10 eV). It is not, therefore, a semiconductor, but electrons and holes can be injected, and their mobilities measured.

7. *Two-dimensional conductors,* particularly those in the inversion layer between a semiconductor and its oxide.

8. Aspects of the behaviour of copper oxide superconductors which depend on disorder.

2 Transport in liquid and amorphous metals; weak-scattering systems

2.1. Introduction

In metals, if the mean free path l is sufficiently large, eqn (1.4) can be used for the conductivity whether the metal is crystalline or not. In this chapter we consider the use of this equation for non-crystalline systems, particularly for liquid and amorphous metals.

Probably the earliest paper dealing with electrons in a field that is not periodic is that of Nordheim (1931) on the resistivity of alloys. For a substitutional impurity in a metal, the potential which scatters the electrons is

$$V_B - V_A$$

where V_A is the potential (or pseudopotential) of the atom of the lattice and V_B of the impurity. This was dramatically shown by the fact that the values of the increase in the resistivity of Cu resulting from 1 per cent of Zn, Ga, and Sn in solid solutions are in the ratio 1, 4, 9 (Linde 1931, 1932a,b; Mott and Jones 1936, p. 293) and thus vary as $(z - 1)^2$, where z is the charge on the atomic core, which is one for copper. Nordheim considered alloys (such as silver–gold) in which the two elements are miscible over the whole range of compositions. If we consider two elements, for which the atomic potentials are V_A, V_B, present in the ratio $1 - x$, x, the average potential is

$$V_{av} = (1 - x)V_A + xV_B.$$

Thus, in each A atom the divergence from this potential is $x(V_B - V_A)$ and in each B atom $(1 - x)(V_B - V_A)$. The total scattering, and hence the resistivity at low temperatures, is thus proportional to

$$\{(1 - x)x^2 + x(1 - x)^2\}\langle U^2 \rangle = x(1 - x)\langle U^2 \rangle$$

where

$$U = \int \psi_{k'}^*(V_B - V_A)\psi_k \, d^3x$$

and $\langle \ \rangle$ denotes an appropriate average over all angles of scattering. In alloy systems such as Ag–Au and Pd–Pt, a variation of the low-temperature resistivity ρ as $x(1 - x)$ is observed.

2.2. Liquid metals

The theory of Nordheim is a weak-scattering theory; it does not consider interference between multiply scattered waves and is therefore valid only if $l \gg a$, where a is the distance between atoms and l the mean free path. The same is true of the theory of the resistivity of liquid metals, first presented by Ziman in 1961 and valid also in the limit $l \gg a$. This work, which made use of the recently developed concept of pseudopotentials, proposed that the scattering potential of each atom in a solid or liquid could be replaced by a *small* pseudopotential. Scattered waves from neighbouring atoms could interfere destructively, but only interference between waves scattered by pairs is considered. Chapter 3 discusses how to go beyond this approximation.

The elements of Ziman's theory are as follows (cf. Faber 1972). First of all, since the scattering is treated as a perturbation and since there is no axis of symmetry, the Fermi surface is taken to be spherical and one starts therefore with the concept of a degenerate electron gas of *free* electrons. Thus the amplitude scattered by two atoms at a distance R from each other is

$$\{1 - \exp(i\mathbf{q} \cdot \mathbf{R})\} f(\theta)$$

where $\mathbf{q} = \mathbf{k} - \mathbf{k}'$, the change in the wave vector on scattering, and $f(\theta)$ is the amplitude scattered by a single atom through an angle θ. Neglecting multiple scattering, the conductivity is then given by (1.4) with

$$\frac{1}{l} = N \int S(q)(1 - \cos \theta)|f(\theta)|^2 2\pi \sin \theta \, d\theta. \tag{2.1}$$

Here N is the number of atoms per cm^3 and $S(q)$ is the structure factor, given by

$$S(q) = N^{-1} \int \{1 + \exp(i\mathbf{q} \cdot \mathbf{R})\}^2 P(R) \, d^3x. \tag{2.2}$$

$P(R)$ is the pair distribution function, $P(R) \, d^3x$ being the probability that another atom is in the volume d^3x at a distance R from a given atom. Using first-order perturbation theory for $f(\theta)$, we find, following Faber and Ziman (1965), for the resistivity

$$\rho = \frac{3\pi}{\hbar e^2 v_F^2 \Omega} \int_0^{2k_F} \frac{|v(\mathbf{q})|^2 S(q) q^3 \, dq}{4k_F^4} \tag{2.3}$$

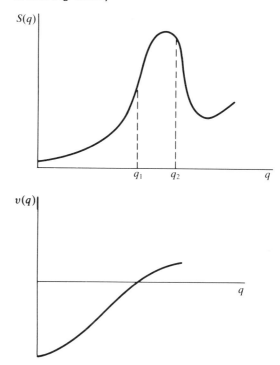

Fig. 2.1. The structure factor $S(q)$ and pseudopotential $v(q)$ for a liquid metal; q_1 and q_2 show the values of $2k_F$ for monovalent and divalent metals.

where

$$v(q) = \int V(r) \exp\{i(\mathbf{q}\cdot\mathbf{r})\}\, d^3x/\Omega,$$

and where v_F and k_F are the values of the velocity and wave vector at the Fermi surface, respectively; the integral is over the volume Ω. Figure 2.1 shows schematically the behaviour of $S(q)$ and $v(q)$. The possibility of applying perturbation theory depends on the assumption that $v(q)$ is small for values of q such that $S(q)$ is large.

One of the most successful applications of the theory is to the temperature dependence of the resistivity of liquid metals. This is large and positive for monovalent metals, small and sometimes negative for divalent metals. This is explained as being caused by the variation with temperature of the structure factor $S(q)$, which can be determined from neutron diffraction. Figure 2.2 shows the results of North *et al.* (1968) for liquid lead at various temperatures. This behaviour is typical. For monovalent metals the resistivity is determined by the left-hand side of the peak. It is observed that the

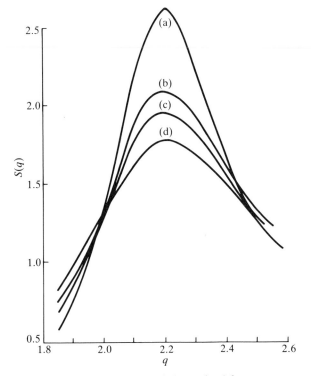

Fig. 2.2. The function $S(q)$ for liquid lead determined from neutron scattering at different temperatures. (a) 340 °C; (b) 600 °C; (c) 780 °C; (d) 1100 °C. (From North *et al.* 1968.)

resistivity of monovalent liquid metals at constant volume is proportional to the absolute temperature; this suggests that $S(q)$ is also proportional to T over the range for which $|v(q)|^2$ is significant. For very low q, the structure factor will be given by the Ornstein–Zernike formula

$$S(q) = k_B T/\beta\Omega$$

where β is the bulk modulus and Ω the atomic volume. This equation describes the contribution from macroscopic fluctuations of density and is valid for either liquids or solids. For divalent metals, on the other hand, a large part of the integral in eqn (2.3) comes from the peak in Fig. 2.2 which shows why for these $d\rho/dT$ can be negative.

The mean free paths in liquid metals can be deduced from the observed resistivity, assuming all valence electrons to be free and that the Fermi surface is spherical. Some values are shown in Table 2.1. Apart from the last two columns in Table 2.1, l is substantially greater than a. One of the first

Table 2.1

Element	Li	Na	Cu	Zn	Hg	Pb	Bi	Te
Valence	1	1	1	2	2	4	5	6
l (Å)	45	157	34	13	7	6	4	0.9

discussions of what happens when $l \sim a$, and of how one gets apparent values less than a, was a treatment of mercury (Mott 1966). Most of the arguments given there are no longer accepted for this material, though we believe they are valid elsewhere. We return to this problem of short means free paths in Chapter 3.

In the Ziman theory, all collisions are treated as elastic; since the atoms are heavy compared with the electrons, the Franck–Condon principle is used, so that the scattering is supposed to be the same as it would be if the atoms were at rest, Greene and Kohn (1965) investigated the error introduced in this way, and found it to be small.

The Hall coefficient R_H in most liquid metals is negative and satisfies the equation

$$R_H = 1/cen,$$

where n is the number of electrons per unit volume. This confirms the hypothesis that the Fermi surface is spherical.

2.3. Mobility of electrons in liquid rare gases

Experiments on the drift mobility of electrons injected into liquid rare gases show high values, which are not thermally activated. Figure 2.3 shows the mobility of electrons in liquid xenon as a function of temperature; the variation is thought to be a consequence of thermal expansion. A theory of this behaviour was given by Cohen and Lekner (1967) and Lekner (1967, 1968). According to these authors the electrons are in a conduction band, and any mobility edge (see p. 25) lies well below $k_B T$ from its bottom. Scattering is caused by thermal fluctuations of the volume. The mean free path is given by

$$1/l = N \int S(q)\lambda^2 (1 - \cos\theta) 2\pi \sin\theta \, d\theta$$

where λ, the scattering length, is

$$\lambda = \Omega_0 m E_0 / 2\pi\hbar^2,$$

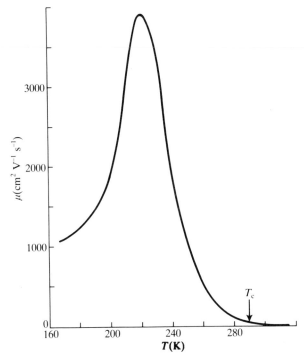

Fig. 2.3. Mobility (μ) of electrons in liquid xenon as a function of temperature. T_c is the critical temperature. The field was 16 V cm^{-1} (Kimura and Freeman 1974).

Ω_0 is the atomic volume, and E_0 is the deformation potential, defined as the change in potential at the bottom of the conduction band for unit expansion. The peak in Fig. 2.3 occurs because E_0 changes sign and therefore goes through a zero as the volume expands.

A discussion by Griko and Popielowski (1977) showed that a treatment along these lines is in general satisfactory, but that, for liquid krypton near the critical point, mobilities are observed some five times higher than given by the theory.

Extensive measurement of the mobility of electrons in other liquids and gases are to be found particularly in the work of Gordon R. Freeman and co-workers in *J. Chem. Phys.* and *Can. J. Phys.* (see for instance Gee and Freeman (1986)).

In all the rare gases except helium and most molecules except H_2 an electron in the conduction band has lower energy than in a vacuum. In a gas for which this is not so, there is the possibility that the electron will make a cavity for itself (Grunberg 1988, 1989). If the difference between vacuum and the bottom of the conduction band is W, the energy to form a cavity of radius r should be

$$(4\pi/3)r^2 S + \hbar^2/mr^2 - W$$

which, choosing r to give the lowest value, namely $r = (\hbar^2/2\pi mS)^{1/5}$, gives a value

$$W_0 = (7/3)\pi^{2/5}S^{2/3}(\hbar^2/m)^{3/5}.$$

If $W > W_0$ a cavity should be formed.

2.4. Amorphous metals and metallic glasses; weak-scattering theory

Amorphous metals can be prepared by many methods, including sputtering, evaporation, chemical deposition, and irradiation of a crystalline solid. Buckel and Hilsch (1954, 1956) did some early work in this field. It was in 1960, however, that Klement, Willencsz, and Duwez first reported that a metallic glass could be formed by the rapid quenching of a metallic alloy from the melt, in this case a liquid alloy of gold and silicon. Since then there has been intensive work on the formation of such metallic glasses. They are true glasses, in the sense that they show a glass transition temperature T_g, with a dramatic rise in the viscosity as the temperature is lowered, together with a specific heat anomaly near T_g. As regards the methods of cooling the material, Duwez first described one frequently used, in which a jet of molten alloy is projected against a cold substrate. As the liquid hits the surface, it flattens out into a thin foil, giving up its heat to the substrate. Instead of a stationary substrate a rotating wheel can be employed; as the molten jet impinges on the wheel, a long thin solid ribbon is thrown off.

Apart from their electrical resistivities, the metallic glasses have interesting mechanical and magnetic properties which will not be discussed here.

Interest in their electrical properties stems from the observation by Mooij (1973) that a negative temperature coefficient of resistance (TCR) is often found in materials for which the resistivity is greater than 150 Ω cm, which as we shall see in § 3.2 is somewhat smaller than the expected (Ioffe–Regel) value. Extensive experimental work on these materials has been carried out and interpreted using the extended Ziman theory (for a review see Naugle 1984). In this theory, the change in the pair distribution function $S(2k_F)$ with temperature should be much smaller than in a liquid, so the temperature coefficient of resistance should be much smaller too. But, following Howson and Gallagher (1988), we do not believe that this theory is applicable to most of these materials, particularly those of high resistivity, which normally contain a transition metal. These authors believe that in this case the resistivity is mainly due to transition from sp electrons to vacant d states,

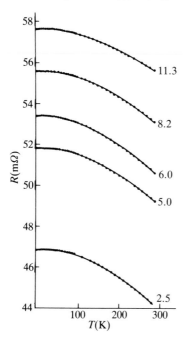

Fig. 2.4. Experimental data for the resistance of $Mg_{70}Zn_{30}$ as a function of temperature at different pressures shown in GPa (Fritsch *et al.* 1982).

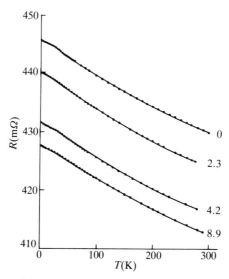

Fig. 2.5. Resistance of $Cu_{57}Zn_{43}$ as a function of T with and without a magnetic field, shown in tesla (Fritsch *et al.* 1982).

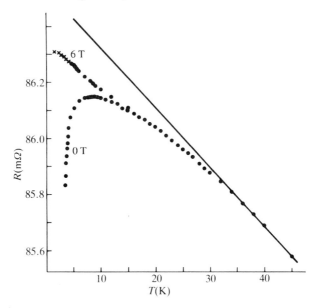

Fig. 2.6. Resistance versus temperature of $Pd_{30}Zr_{70}$ in two magnetic fields (values in tesla). The solid line is an extrapolation of the high-temperature results (Willer *et al.* 1983).

for which the density of states $N_d(E_F)$ is high. Mott (1936) first described the electrical resistance of transition metals in this way, the time of relaxation for the conduction electrons being given by

$$1/\tau \propto N(E_F).$$

For crystalline materials the transition is caused by interaction with phonons—but probably here disorder is responsible.

Any theory of the TCR which depends on interaction with phonons must give a value of $\rho^{-1} \, d\rho/dT$ which tends to zero as $T \to 0$. Some materials, for instance $Mh_{70}Zn_{30}$ (Fritsch *et al.* 1982), do indeed show the behaviour illustrated in Fig. 2.4, but for others the negative TCR remains constant down to ~ 2 K. For instance, Fritsch *et al.* (1982) and Willer *et al.* (1983) investigated $Ti_{50}Be_{40}Zr_{10}$; their results are shown in Fig. 2.5. Some alloys are superconducting and show this effect only in a magnetic field; an example is shown in Fig. 2.6 (from Willer *et al.* 1983). Such effects must be the result either of multiple scattering or of electron–electron interaction (Imry 1980); they are discussed in Chapters 3 and 5. A further reason why the Ziman theory is not adequate is that the phenomenon of quantum interference occurs, being responsible for a $T^{1/2}$ behaviour of $\Delta\rho$. For this reason we return to the subject in Chapter 5.

3   Short mean free paths and Anderson localization

3.1. The Kubo–Greenwood formula

A main theme of this book is the behaviour of the conductivity of metals and semiconductors when the mean free path k is of order a, the distance between atoms or scattering centres. In this case the assumption of single scattering on which the Boltzmann formula (eqn (1.4)) is based breaks down. One way of treating this problem is by using the Kubo–Greenwood formula (Kubo 1956; Greenwood 1958) which we shall now derive. In this formula the conductivity $\sigma(\omega)$ at frequency ω is calculated, and ω is then allowed to tend to zero. We suppose that an electric field $F \cos \omega t$ acts on a specimen of volume Ω. Then the chance P per unit time that an electron makes a transition from a state with energy E to any of the (degenerate) states with energy $E + \hbar\omega$ is

$$P = \tfrac{1}{4}e^2 F^2 (2\pi/\hbar)|\langle E + \hbar\omega|x|E\rangle|^2 \Omega N(E + \hbar\omega). \qquad (3.1)$$

The matrix element is defined as

$$\langle E'|x|E\rangle = \int \psi_{E'}^* x \psi_E \, \mathrm{d}^3 x,$$

the functions ψ_E being normalized to the volume Ω and E' denoting any of the (degenerate) states $E + \hbar\omega$. It is convenient to write

$$\langle E + \hbar\omega|x|E\rangle = (\hbar/m\omega)D_{E+\hbar\omega, E}$$

where

$$D_{E', E} = \int \psi_{E'}^* (\partial/\partial x)\psi_E \, \mathrm{d}^3 x.$$

That these are equal can be shown directly, or by remembering that the optical transition probability can be represented by the square of the matrix element of Eex, where E is the strength of the field, or by using the vector potential A $(=cE/\omega)$, where the perturbing potential energy is taken as

$$\frac{e\hbar}{mci} A \frac{\partial}{\partial x}.$$

Both must lead to the same result. Equation (3.1) then becomes

$$P = (\pi e^2 \hbar \Omega / 2m^2 \omega^2) F^2 |D|^2 N(E + \hbar\omega). \tag{3.2}$$

The quantity $\frac{1}{2}\sigma(\omega)F^2$ is the mean rate of loss of energy per unit volume. To obtain this, we must multiply (3.2) by the following factors:

1. $N(E)f(E)\,dE$, the number of occupied states in the range dE; $f(E)$ is here the Fermi distribution function.
2. $1 - f(E + \hbar\omega)$, the chance that the state with energy $E + \hbar\omega$ is unoccupied.
3. $\hbar\omega$, the energy absorbed in each quantum jump.
4. The factor 2 for the two spin directions.

We find, integrating over all energies,

$$\sigma(\omega) = \frac{2\pi e^2 \hbar^2 \Omega}{m^2 \omega} \int [f(E)\{1 - f(E + \hbar\omega)\} - f(E + \hbar\omega)\{1 - f(E)\}]$$

$$\times |D|^2_{\text{av}} N(E)N(E + \hbar\omega)\,dE. \tag{3.3}$$

The second term in the square brackets gives the energy emitted in downward jumps. The quantity in square brackets simplifies to

$$f(E) - f(E + \hbar\omega),$$

so (3.3) reduces to

$$\sigma(\omega) = \frac{2\pi e^2 \hbar^3 \Omega}{m^2} \int \frac{\{f(E) - f(E + \hbar\omega)\}|D|^2_{\text{av}} N(E)N(E + \hbar\omega)}{\hbar\omega}\,dE.$$

When the temperature is zero this becomes

$$\sigma(\omega) = \frac{2\pi e^2 \hbar^3 \Omega}{m^2} \int \frac{|D|^2_{\text{av}} N(E)N(E + \hbar\omega)}{\hbar\omega}\,dE. \tag{3.4}$$

For the d.c. conductivity we make ω tend to zero and obtain

$$\sigma_E(0) = \frac{2\pi e^2 \hbar^2 \Omega}{m^2} |D_E|^2_{\text{av}} \{N(E)\}^2 \tag{3.5}$$

where

$$D_E = \int \psi^*_{E'} \frac{\partial}{\partial x} \psi_E\,d^3x, \quad E = E'.$$

and the subscript av represents an average over all states with energies E.

For long mean free paths it must be possible to reduce this formula to the Boltzmann form (1.4). This was first done by Edwards (1961); proofs are given in Mott and Davis (1979, p. 13). An approximate derivation may be given as follows. We may represent the integral D by the sum of volumes l^3 in which the wave functions are coherent. These will have random signs, so

$$D = (\Omega/l^3)^{1/2}\delta,$$

where

$$\delta = k \int^{l^3} \exp\{i(\mathbf{k}' - \mathbf{k}) \cdot \mathbf{r}\} \, d^3x/\Omega.$$

Setting $|\mathbf{k} - \mathbf{k}'| = 2k \sin \frac{1}{2}\theta \simeq k\theta$, where θ is the angle between the two values of $\mathbf{k}$, we approximate by writing

$$\delta = kl^3/\Omega, \quad \text{if} \quad kl\theta < 1$$
$$= 0, \qquad \text{otherwise.}$$

Thus

$$|D|^2_{\text{av}} = (\Omega/l^3)k^2l^6\Omega \int^{1/kl} 2\pi\theta \, d\theta/4\pi$$

$$= \pi l/3\Omega.$$

Substituting for $N(E)$ in the form

$$N(E) = km/2\pi^2\hbar^2,$$

we find that

$$\sigma = \text{const } e^2k^2l/\hbar;$$

the correct value of the constant cannot be obtained in this way.

The Kubo–Greenwood formula, then, gives new results only for small values of the mean free path; these will be discussed below.

3.2. Anderson localization and the mobility edge

Anderson's paper of 1958 on 'The absence of diffusion in certain random lattices' has had a profound effect on our understanding of the behaviour of electrons in non-crystalline media, introducing as it did the concept of 'localization'. Anderson considered a crystalline array of potential wells, as illustrated in Fig. 3.1(a). The tight binding approximation is used; that is to say, overlap between wave functions is considered negligible except for wells that are nearest neighbours. Then the approximate solutions Ψ of the Schrödinger equation (1.1) are

$$\Psi = \sum_n \exp(i\mathbf{k} \cdot \mathbf{a}_n)\psi(|\mathbf{r} - \mathbf{a}_n|).$$

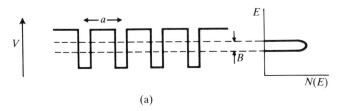

(a)

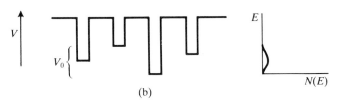

(b)

Fig. 3.1. (a) Potential wells for a crystalline lattice. (b) The same with a random potential energy at each well, as in Anderson (1958). The density of states $N(E)$ is also shown.

Here **k** is a wave vector, $\mathbf{a}_n$ are the lattice points, and $\psi(\mathbf{r})$ is the wave function for an electron in any of the wells, supposed spherically symmetrical (s-like). The states then lie in a band of width

$$B = 2zI$$

where z is the coordination number and I is the transfer integral

$$I = \int \psi(|\mathbf{r} - \mathbf{a}_n|)H\psi(|\mathbf{r} - \mathbf{a}_{n+1}|)\, \mathrm{d}^3x.$$

H is here the Hamiltonian. For hydrogen-like functions we write $I = I_0\, \mathrm{e}^{-\alpha R}$ where

$$I_0 = (e^2\alpha/\kappa)(1 + \alpha R).$$

Figure 3.2 shows the calculated density of states for a simple cubic structure ($z = 6$) in the tight binding approximation used here.

Anderson introduced a random potential at each well, the depths lying in the range V_0. If $V_0 \ll B$, the effect will be to introduce a mean free path l, and, using the Born approximation, it can be shown that

$$a/l = 0.7(V_0/B)^2.$$

If then $V_0 \simeq B$, we see that $l \sim a$. This means that the wave function loses

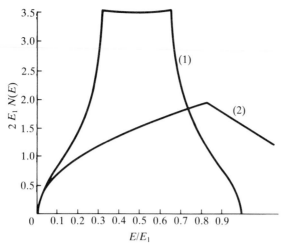

Fig. 3.2. Density of states for (1) a simple cubic lattice according to the tight bonding approximation; (2) for free electrons. E_1 is the bandwidth.

phase coherence in passing from one well to the next, so that it takes the form

$$\Psi = \sum_n c_n \exp(i\phi_n)\psi(|\mathbf{r} - \mathbf{a}_n|). \tag{3.6}$$

Here the ϕ_n are random phases and the c_n are constants. Equation (3.6) has been called the random phase approximation. It is obvious that the wave function cannot lose phase memory faster than this, and according to a principle introduced by Ioffe and Regel (1960) this means that the mean free path cannot be less than a. According to eqn (1.4), the conductivity when $l = a$ should be, since $S_F = 4\pi k_F^2$ and $(4\pi/3)k_F^3 = (1/2a^3)8\pi^3$ (whence $k_F a \simeq \pi$),

$$\sigma_{IR} = ce^2/3\hbar a \tag{3.7}$$

where $c = (3/\pi)^{2/3} \simeq 1.0$. We call this the Ioffe–Regel value of the conductivity. If $a = 3$ Å, $\sigma_{IR} = 700\ \Omega^{-1}\ \text{cm}^{-1}$, corresponding to $\rho = 1.3 \times 10^3\ \mu\Omega\ \text{cm}^{-1}$. In some cases, for instance in some transition metals, the conductivity, with increasing temperature, saturates at near this value.

Equation (3.7) has to be corrected in two ways. First, since l is small, multiple scattering may not be negligible, Kawabata (1981), Larkin and Khmelnitzkii (1979), and Shapiro and Abrahams (1981) in different forms gave the equation

$$\sigma = \sigma_B\{1 - C/(k_F l)^2\} \tag{3.8}$$

to take account of this. They obtained this form by summing the appropriate diagrams. C is a constant which is not precisely known but is of order 1. σ_B

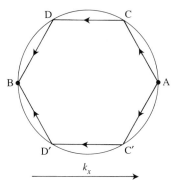

Fig. 3.3. Model of Bergmann for quantum interference, ACDB and AC'D'B are equivalent paths in k-space for scattering, leading to interference berween them.

is the Boltzmann conductivity (1.4). Kaveh and Mott (1983) first pointed out that (3.8) should be generalized to the form

$$\sigma = \sigma_B\left\{1 - \frac{C}{(k_F l)^2}\left(1 - \frac{l}{L}\right)\right\} \qquad (3.9)$$

where L is either the size of the specimen, or the inelastic diffusion length L_i, that is the distance an electron will diffuse before an inelastic collision, or, in a magnetic field H, the cyclotron radius $(c\hbar/eH)^{1/2}$. This equation has the surprising result that inelastic scattering, such as scattering by phonons, can *increase* the conductivity. This is because the constructive interference between the scattered waves, which increases the scattering and reduces the conductivity, is reduced if the volume available (L^3 or L_i^3) is small.

The simplest way to understand the origins of eqn (3.9) is by using the model of Bergmann (1983, 1984), though his work was for a two-dimensional system. Figure 3.3 shows the Fermi surface in k-space for a degenerate gas of free electrons. An electron with momentum represented by the point A can be scattered by a multiple scattering process to the point B by two identical paths such as ACDB and AC'D'B. If the scattering is elastic, the amplitudes at B instead of the intensities must be added. The waves will be in phase over an area (in three dimensions) const/$(kl)^2$, leading to the reduction in the conductivity shown in eqn (3.8). The term $(1 - l/L)$ represents the reduction in the scattered term if not all collisions are elastic. L has been defined above. Also for an alternating field of frequency ω, $L = L_\omega = (D/\omega)^{1/1}$ where D is the diffusion coefficient $\sigma/e^2N(E)$ (Ortuno and Kaveh 1984, Kawabata 1984). If the different lengths L are comparable in magnitude, we take

$$\frac{1}{L_{av}^4} = \frac{1}{L_{L_i}^4} + \frac{1}{L_H^4} + \cdots.$$

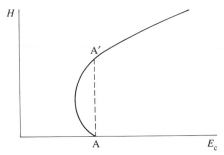

Fig. 3.4. Suggested position of a mobility edge E_c; if E_c lies at A in the absence of a magnetic field, increasing the field gives two transitions.

Although these equations are derived on the assumption that the correcting term in (3.8, 3.9) is small, we use it to extrapolate to the transition; Kaveh and Mott (1987) give some arguments to suggest that this is a valid approximation. Also, we assume that multiple scattering, without quantum interference can be included in the formula for σ_{IR} given by (3.7) without too great an error, σ_{IR} in any case being only approximately defined.

In the case of a magnetic field, H, when $L_H = (\hbar c/eH)^{1/2}$, we see that the field may first produce a negative magnetoresistance by the destruction of quantum interference, followed by a positive term resultant on the shrinking of the orbits in directions perpendicular to the field. Shapiro (1984) pointed out the mobility edge could behave, as a function of H, as in Fig. 3.4.

The inelastic diffusion length, L, is given by the equation

$$L_i = (D\tau_i)^{1/2}$$

where τ_i is the time leading to an inelastic collision and D is the diffusion coefficient. Since D is related to the conductivity by the equation

$$\sigma = e^2 N(E_F)D,$$

D will to a first approximation be independent of temperature. If τ_i is determined by collision with other electrons, on the other hand, we expect, since the Landau–Baber theory gives for a pure metal $\rho \propto T^2$, that

$$\tau_i \propto \tau_0 (E_F/k_B T)^2,$$

where τ_0 is a time of order $\hbar/E_F$. Thus L_i should be proportional to $1/T$, and from (3.9) a negative temperature coefficient of resistance is expected, proportional to T. If on the other hand τ_i is determined by collisions with phonons, τ_i is proportional to $1/T$ and the negative temperature coefficient of resistance goes as $T^{1/2}$. Both kinds of behaviour have been observed in amorphous metals (Chapter 5).

The second correction to eqns (3.7) and (3.9) is the following. If V_0 in Fig.

3.1 exceeds B, the density of states is reduced, as Fig. 3.1 shows. We may denote by g the drop in the density in mid-band produced by disorder. This has been written (Mott 1974)

$$g = B/\{1.74\sqrt{(V_0^2 + B^2)}\}. \tag{3.10}$$

The factor 1.74 may be understood from Fig. 3.2; for the crystal the calculated density in mid-band exceeds by the factor 1.74 the quantity $1/Ba^3$. Then eqn (3.5) shows that the conductivity without the Kawabata correction (3.9) should be decreased by g^2, so that

$$\sigma = \tfrac{1}{3}e^2g^2/\hbar a. \tag{3.11}$$

In liquids, g can be determined from the Knight shift or paramagnetic susceptibility, and evidence for eqn (3.11) will be shown in Chapter 8. In liquids, too, all collisions are inelastic and it is suggested there that $l = L_i$, so the correction in (3.9) does not apply. In solids, however, the correcting term is also affected by g and we can write (Mott and Kaveh 1985a, b)

$$\sigma = \sigma_B g^2 \left\{ 1 - \frac{C}{g^2(k_F l)^2}\left(1 - \frac{l}{L}\right)\right\}. \tag{3.12}$$

It is important to emphasize that only in the case when the mean free path has its smallest value ($l = a$) is the factor g^2 appropriate in the conductivity. While we may still write $\sigma = S_F e^2 l g^2/12\pi^3\hbar$ when $l > a$, it can then be shown that $l = l_z/g^2$, where l_z is the value calculated by perturbation methods (Edwards 1961); using l_z the term g^2 therefore cancels out.

If V_0 increases, then g decreases and according to eqn (3.11) so does the conductivity. As already suggested, we make the assumption that reasonable results can be obtained by extrapolating (3.1)—or other similar but more complicated equations to be obtained later—to the transition where the conductivity vanishes. Kaveh and Mott (1987) give an argument for this which will be described in the next section. In addition, the assumption gives results in agreement with Anderson's criterion for localization, as we show in the following sections.

In this book, then, we shall use eqn (3.12) to describe what happens at the Anderson metal–insulator transition. The main point of Anderson's paper was that a certain value of V_0/B, the wave functions become localized. Calculations of the exact value are difficult. For a coordination number $z = 6$ and an energy at mid-band, Anderson (1958) gave 5, Edwards and Thouless (1972) gave 2, and the most recent numerical work (Elyutin et al. 1984) gave 1.7, which is the value used in this book and in fair agreement with other recent calculations (see Mott 1991, p. 36). Early work by Weaire and Srivastava (1977) and co-workers illuminates the reason for the differences between the results of earlier work. The factor g at the transition is then 0.3.

By 'localized' we mean that the wave functions are no longer of the form (3.6) but rather that they describe electrons in traps, and take the form of the real part of

$$\Psi = \exp\{-(\mathbf{r} - \mathbf{r}_0|)/\xi\} \sum_n C_n \exp(i\phi_n)\psi(|\mathbf{r} - \mathbf{r}_0|) \qquad (3.13)$$

each one being localized around a point r_0 in space. ξ is supposed to be infinite when localization first occurs and then to decrease with increasing V_0. It is called the localization length.

If we take a half full band and if $l = a$, then $k_F l = \pi$. For g we use eqn (3.10), so using the result of Elyutin et al. at the transition we find $g \simeq 1/3$. Therefore (3.12), extrapolated to the transition, gives a result for the transition to a non-metallic state in agreement with that predicted by these authors from Anderson (1958).

If states are localized at the Fermi energy, the conductivity at zero temperature vanishes. Conduction can occur by the thermally activated movement from one localized state to another, a process called hopping and first described by Miller and Abrahams (1960) and discussed in § 3.5. For some years after the publication of Anderson's paper, his result was widely queried because it was thought that an electron, by tunnelling far enough, could find a state with the same energy. Though this is so, it is found that on taking an ensemble average this does not affect the insulating property, though the activation energy for conduction does tend to zero with decreasing temperature. So an insulating state for a degenerate electron gas with a finite density of states at the Fermi energy can exist. A material in such a state has been called a 'Fermi glass'.

If the parameter V_0/B is not great enough to give localization throughout the band, Mott (1967) first pointed out that states in the band tails would be localized and that energies of localized and extended (non-localized) states would be separated by a sharp energy E_c, known (Cohen et al. 1969) as a 'mobility edge'. The lower extremity of such a band is shown in Fig. 3.5, the range of energies in which states are localized being shaded. The energy interval $E_c - E_A$ is not easy to calculate; for the Anderson model a value has been given by Abou-Chacra and Thouless (1974), for the conduction band of amorphous silicon by Davies (1980). For the conduction band of a non-metal, the existence of a mobility edge means that the lowest states have become traps, and conduction will normally be by electrons excited to mobility edge E_c, so that the conductivity behaves as

$$\sigma = \sigma_0 \exp\{-(E_c - E_F)/k_B T\}. \qquad (3.14)$$

For a degenerate electron gas, metallic behaviour is expected if the Fermi energy E_F lies above E_c, as is shown in Fig. 3.5; if it lies below E_c, the material

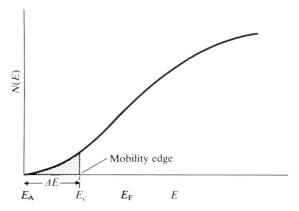

Fig. 3.5. Showing the conduction band in a non-crystalline material; E_c is the mobility edge and energies for which states are localized are shaded.

is an insulator and conduction is either by hopping (§ 3.5) or by excitation of carriers to E_c.

The conductivity of the electron gas when E_F lies at E_c and the pre-exponential factor σ_0 are identical. This is because both may be written $eN(E_F)/k_B T\mu$, where μ is the mobility for an electron with energy at E_c. The value of this quantity is of considerable interest. Before the existence of the correcting term in (3.12) was realized, I argued that because the factor g^2 would have the value at the mobility edge, namely

$$1/(1.74)^2\{1 + (1.7)^2\} = 1/11.5 = 0.085,$$

σ_0 would be given by

$$\sigma = Ce^2/\hbar a \tag{3.15}$$

where $C = 0.33 \times 0.085 \simeq 0.03$.

This value of σ was designated the 'minimum metallic conductivity' (σ_{min}) it being thought that *metallic* conductivity could not drop below this value (Mott 1975). For $a = 3$ Å it is $2 \times 10^2 \ \Omega \ cm^{-1}$, but for impurity bands for which $a \sim 30$ Å, it is of course much smaller. It was argued that the conductivity at absolute zero would reach this value as the disorder increased, and then fall discontinuously to zero.

The scaling theory put forward by Abrahams *et al.* (1979), which will be described in § 3.4, predicted on the other hand that at zero temperature σ would drop continuously to zero. If then as we assume eqn (3.13) is correct over the whole range, with $C = 1$ and $L = \infty$ as at zero temperature, this is clearly the case. Since $k_F L \sim \pi$ in the Ioffe–Regel limit, for one electron per centre the transition will occur when $g \simeq \frac{1}{3}$, the value obtained above from the calculation of Elyutin *et al.* (1985). Evidence that this is so is reviewed later. Thus the quantity σ_{min} does not have any clear significance for solids at zero temperature, except for certain cases in a strong magnetic field (§ 3.11). It will be seen, however, that if L_i is finite, σ drops to

$$\sigma_{min} l/L_i \qquad (3.16)$$

In liquids probably all collisions are inelastic, the second term in (3.12) is then absent, and σ_{min} is both a minimum metallic conductivity and the pre-exponential factor σ_0 for a semiconductor. This is discussed further in Chapter 8.

We now consider the behaviour of the wave function and the conductivity for electrons with energies near E_c. Below E_c the functions are localized. As always with the Anderson model of Fig. 3.1, they may be written in the form (3.13); ξ is believed to vary with E as

$$1/\xi = \text{const}(E_c - E)^s \qquad (3.17)$$

where s is thought to be equal to unity. A theoretical demonstration that this is so was given by Mott (1984), and is also set out in Mott (1991, p. 42). At an energy ΔE above E_c, Mott (1984) and Mott and Kaveh (1985a,b) argued that the form of the wave functions is not greatly changed, and therefore that

$$\sigma \sim \sigma_{min} a/\xi \qquad (3.18)$$

where ξ is the localization length at an energy ΔE below E_c. If (3.17) and (3.18) are valid, the conductivity of a degenerate electron gas should go to zero linearly with $E_F - E_c$, if this quantity is varied. We shall see in Chapter 4 that this is so for some systems, but for others a behaviour as $(E_F - E_c)^{1/2}$ is observed. An explanation of why this should be so was given by Kaveh (1985) and is discussed in § 5.3 of this book.

It is often stated in the literature that Anderson localization occurs when l attains the Ioffe–Regel value. The reverse is true, but the disorder at which Anderson localization occurs is much greater than that for Ioffe–Regel, and indeed greater than that for $\sigma = \sigma_{min}$.

3.3. The localization length for energies above E_c

We have already introduced the concept of the localization length ξ (eqn (3.12)). In materials in which a mobility edge E_c separates the energies of localized from non-localized states, ξ should tend to infinity as $E_c - E$ tends to zero. We therefore write

$$\xi = a\{E/(E_c - E)\}^\nu. \qquad (3.19)$$

We believe that $v = 1$ and shall give arguments to show this, though numerical work by Schreiber, Kramer, and Mackinnin (1989) give $v = 1.6$. We do not believe that, in the many-body problem (a condensed gas), interactions affect the value of v.

Mott (1984) and Mott and Kaveh (1985a,b) have given the following description of the behaviour of the wave function for energies near E_c. Since the Kubo–Greenwood analysis shows that σ is given by the sum of squared quantities, one has to understand how σ can go continuously to zero (so that there is no minimum metallic conductivity). According to Mott (1984), this is to be understood by considering the form of the wave function ψ for an electron with energy ΔE above E_c. Over a range in space of length ξ it will differ little from the localized states at a small energy ΔE below it. ξ is defined by the length occupied by ψ for these localized states below it (see eqn (3.19)). The function for the extended state is therefore as in Fig. 3.6. In the peaks the wave function is real, but it is complex in the neighbourhood of the minima. The Kubo–Greenwood integral (3.4) is therefore of the form

$$\int \psi_1^*(\partial/\partial x)\psi_2 \, dx$$

and over the range x the ψ_1 and ψ_2 are real and identical, so the integral takes the form $[\psi^2]_{x_1}^{x_2}$, where x_1, x_2 lie in adjacent minima. It will be seen why σ tends to zero as E tends to E_c, and how the localization length above E_c can be defined. Mott (1984) gives a further discussion to show that $\sigma = 0.03e^2/\hbar\xi$.

His argument is as follows. We consider states with energies lying between E_c and some lower energy E such that

$$\Delta E = E_c - E$$

is small. Wave functions will be localized, normally non-degenerate, and therefore, in the absence of a magnetic field, real. We suppose them to be of

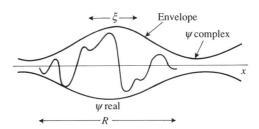

Fig. 3.6. Wave functions for energies just above E_c according to Mott (1983). ξ is the localization length for a function at the same energy below E_c.

the form

$$Re(\psi_{ext}) \exp(-r/\xi)$$

where ξ depends on ΔE as

$$\xi \sim a(\Delta E/E)^s.$$

We take $s = 1$, though our results do not depend critically on this value.

Two wavefunctions ψ_1, ψ_2 will split into $\psi_1 \pm \psi_2$ and so cannot overlap unless the transfer integral $\int \psi_1 H \psi_2 \, d^3x$ is less than ΔE. For two states with comparable values of ζ, the transfer integral will be of the form

$$H_0 \exp(-R/\xi)$$

so two states cannot be closer in energy than at a distance R given by

$$R = \xi \ln(H_0/\Delta E) = \xi \ln(aH_0/E\xi). \tag{3.20}$$

We think that, in the region where ψ_1 overlaps ψ_2, it must be identical with ψ_2; so we do not expect a term $(a/\xi)^{3/2}$ from the random phases, as we would in the case far from the mobility edge.

H_0/E should be of order unity. We also argue that, although below E_c there will be functions with ξ much larger than R in (3.19), there will be very few of them, because of their large overlap.

The essential point is that just *above* the mobility edge, the functions will be little changed within a volume R^3. Thus:

1. Within this volume, the functions ψ_1, ψ_2 remain real and identical with (3.13), as we have seen.

2. The functions ψ_1, ψ_2, etc. have long-range fluctuations, of wavelength R given by (3.20), so that the quantity

$$\left\{ V \int |\psi|^4 \, d^3x \right\}^{-1}$$

tends to zero; V is here the volume. This is shown by other investigators (see, for example Aoki 1983). Since for all the functions ψ the maxima are in the same positions in space, namely that given by the function below E_c, and depends on the fourth power of ψ, this property should multiply σ by the factor

$$\{\exp(R/\xi)\}^2.$$

The constant C_n in (3.13), though fluctuating from site to site, will decrease steadily in the range $\xi < r < R$. In previous work we have supposed that these fluctuations have little effect on σ, though this may be doubted in view of the 'fractal' property revealed by the calculations of various authors (e.g. Zollen 1983, Chap. 4).

3. However, from the argument (1) we see that

$$\int_{R^3} \psi_1^*(\partial/\partial x)\psi_2 \, dx \, dy \, dz = \frac{1}{2}\int dy \, dz [\psi^2]_x^{x+R}$$

apart from some constant factor $\exp(i\varphi)$. x and $x + R$ are supposed to be values at the minima of ψ. Here the phases are random, so that both for the yz integration and for the addition of the next term resulting from the x integration, we can use the random-phase approximation. So our conclusion is that $\psi_1^* = \psi_2$ which reduces the integral by a term $(a/\xi)^{1/2}$, the range of r from $r = \frac{1}{2}\xi$ to $r = \frac{1}{2}R$ contributing little to the integral.

Since ψ^2 must be taken at a distance of $\frac{1}{2}R$ from the maxima, from the x integration a term is introduced which just cancels. Thus near E_c,

$$\sigma = 0.03 \frac{e^2}{\hbar a} \frac{a}{\xi} = 0.03 \frac{e^2}{\hbar\xi}. \tag{3.21}$$

Our value 0.03 is near that of $\frac{1}{2}\pi^2$ given by Vollhardt and Wölfle (1980a,b).

3.4. Can localized and extended states coexist at the same energies?

Several authors have suggested that localized and extended states can coexist in the same small energy range, and thus, for an infinite specimen for which the separation between the energies is zero, for the same energy. Srivastava (1989, 1990) considers that when this occurs two localized states result but with abnormal properties. Phillips (1983) claims to have proved that these can coexist, and has used the concept recently (Phillips 1990) to explain certain properties of high-T_c superconductors. Except for a negligible number of states at special energies, the present author does not believe this to be true. Reasons are as follows.

Let us consider a three-dimensional system of linear size L; we shall allow L to tend to infinity. In this limit suppose that a comparable number of localized states ϕ_1 and extended states ϕ_2 exist in the same small energy range. All are supposed normalized, and since they are eigenstates they are orthogonal. Thus the amplitude of $|\phi_2|^2$ for all extended states tends to zero as $1/L^3$, while that of $|\phi_1|^2$ is of order $1/\xi^3$.

Consider now the function

$$\psi = A_1\phi_1 + A_2\phi_2$$

where ϕ_1, ϕ_2 are adjacent localized and extended states separated by an energy $\sim \hbar^2/mL^2$. This of course is not a solution of Schrödinger's equation, except in the limit $L \to \infty$. In this limit, however, it is. It is extended, and has the property that, whereas in most of space if we normalize ψ to one particle per unit volume, there exists a volume in which it diverges as $(L/a)^3$. Moreover, if the proportion of localized states is significant (say a finite fraction η of the whole), then in any volume ξ^3/η, if we pick the right energy, we should find that a solution of the Schrödinger equation exists which decays exponentially to zero and yet, when $|\psi|^2$ is integrated over all space, the integral is comparable with that within a few multiples of ξ from the centre of the localized region.

While we have no formal proof that such states cannot exist, it seems so improbable as to raise doubts about the possibility of coexistence.

3.5. Hopping conduction

If the Fermi energy E_F lies below the mobility edge E_c, we have seen that conduction may be of two kinds.

1. By excitation to the mobility edge. We may then give σ_0 in eqn (3.14) the value

$$\sigma_0 \simeq 0.03e^2/\hbar L_i.$$

The inelastic diffusion length may then be the result of collisions with phonons, or Auger processes in which an electron loses energy to another which has energy below E_F.

2. By thermally activated hopping, if $N(E_F)$ is finite. This is a process in which an electron in an occupied state with energy below E_F receives energy from a phonon, which enables it to move to a nearby state above E_F. A process of this kind was first described by Miller and Abrahams (1960) as an explanation of impurity conduction in doped and compensated semiconductors (Chapter 4). In this work, the electron was supposed always to move to the nearest empty centre. Their analysis resulted in an expression for the conductivity

$$\sigma = \sigma_3 \exp(-\varepsilon_3/k_B T).$$

ε_3 is expected to be of the form

$$\varepsilon_3 \sim 1/N(E_F)a^3$$

where a is the distance between nearest neighbours. This is discussed further in Chapter 4.

Mott (1968) first pointed out that at low temperatures the most frequent hopping process would *not* be to a nearest neighbour. The argument in its simplest form is the following. Within a range R of a given site the density of states per unit energy range is, near the Fermi energy,

$$(4\pi/3)R^3 N(E_F).$$

Thus for the hopping process through a distance R with lowest activation energy, this energy ΔE will be the reciprocal of this,

$$\Delta E = 1/(4\pi/3)R^3 N(E_F).$$

Thus, so far as the activation energy is concerned, the further the electron hops the smaller will be ΔE. But hopping over a large distance involves tunnelling and the probability will contain a factor

$$\exp(-2\alpha R)$$

where $1/\alpha$ is the decay length of the localized wave function. So there will be an optimum hopping distance R, for which

$$\exp(-2\alpha R)\exp(-\Delta E/k_B T)$$

is a maximum. This will occur when

$$2\alpha R + 1/\{(4\pi/3)R^3 N(E)k_B T\} \tag{3.22}$$

has its minimum value, that is when

$$R = \{1/8\pi N(E)\alpha k_B T\}^{1/4} \tag{3.23}$$

Substituting for R in (3.22), we see that the hopping probability and thus the conductivity is of the form

$$A \exp(-B/T^{1/4}), \tag{3.24a}$$

where

$$B = 2\left(\frac{3}{2\pi}\right)^{1/4}\left(\frac{\alpha^3}{k_B N(E_F)}\right)^{1/4}. \tag{3.24b}$$

For other methods of deriving this equation, giving somewhat different values of B, see Mott and Davis (1979, p. 32) or Shklovskii and Efros (1984). In two-dimensional problems, 1/3 replaces 1/4 (Hamilton 1972).

This form of conduction is called 'variable-range hopping'. On the experimental side, both in doped crystalline semiconductors and amorphous materials it has frequently been observed, and the form

$$\sigma = A \exp(-B/T^\nu) \tag{3.25}$$

often represents the behaviour. Experimentally, however, it is difficult to determine the value of ν.

There is an extensive literature on the value of the constant A. A review giving values for single and multiphonon hopping is given by Emin (1975).

For a recent discussion see Summerfield and Butcher (1985) and Schirmacher (1991).

Efros and Shklovskii (1975) first proposed that, when the Coulomb interaction between the electrons is taken into account, the index v in (3.25) should be given by $v = \frac{1}{2}$, their analysis being confirmed by calculations of J. H. Davies *et al.* (1982). They obtained this result as follows. They considered an empty and an occupied state at a distance R from each other with energies ε_a, ε_b above and below the Fermi level. The energy required to move an electron from one to the other is

$$\varepsilon - e^2/\kappa R$$

where $\varepsilon = \varepsilon_a - \varepsilon_b$. Thus around the volume occupied by any one of the states there is a sphere of volume

$$(4\pi/3)R^3 = (4\pi/3)(e^2/\kappa\varepsilon)^3$$

in which the other electron cannot be located.

To proceed further we must consider the several approximations to the density of states. $N(E)$ is the density of states calculated for some averaged field (Hartree or Hartree–Fock) produced by all the other electrons. This (Thouless 1970) is finite at E_c, as are its derivatives. We call it the density of states for non-interacting particles. Next we denote by $N_0(E)$ the density of states seen by an electron introduced or taken out of the system, if the surrounding electrons are not allowed to relax. We call this the single-particle density of states, and it is assumed that for hopping conduction, as for tunnelling, this is the correct form to use. And finally we have the thermodynamic density of states $\tilde{N}(E)$, where all the particles *are* allowed to relax, which must be correct for the specfic heat and paramagnetism.

Efros and Shklovskii argued that the number of states excluded within a range of energies distant ε from the Fermi level is

$$(4\pi/3)(e^2/\kappa e)^3 n(\varepsilon)$$

where

$$n(\varepsilon) = \int_0^\varepsilon N_0(E)\, \mathrm{d}E.$$

This must not tend to infinity as $\varepsilon \to 0$, so $n(\varepsilon)$ must vary as ε^3 or a higher power of ε and $N_0(E)$ as $(E - E_F)^s$ with $s \geqslant 2$. Computer studies give $s = 2$ in most cases. The density of states is thus expected to have a parabolic minimum. So $N(E)$ in (3.24b) should be replaced by $(e^2/\kappa a)^2\alpha^{-3}$ where $\alpha = 1/\xi$, yielding

$$\sigma = \sigma_0 \exp\{-(T_0/T)^{1/2}\}, \tag{3.26}$$

where $k_B T_0 = $ const $e^2/\kappa a$; according to Shklovskii and Efros (1984) the constant is ~ 2.8.

We believe that this form should be correct for impurity conduction, i.e. concentrations far from the Anderson transition, so that $\alpha a \ll 1$, (which means that the localization length is small compared with the distance between centres). If not, the argument above breaks down, and the index should be 1/4, as in Mott's original theory. There is much experimental evidence that a change occurs from $v = 1/4$ to $v = 1/2$ in impurity conduction as the concentration of dopant moves away from that at which localization occurs (Shafarman and Castner 1986; Shafarman *et al.* 1986; Mott 1991).

The key is that of Shafarman *et al.* (1989). They find the index 1/4 when $\langle E(T) \rangle$ is greater than the Coulomb gap ($E(T)$ is the hopping energy) and this leads to $T > T_0/1170$ where

$$kT_0 = 18/\xi^3 N(E_F) \tag{3.27}$$

and $v = 1/2$ when T is well below $T_0/2000$.

Other papers by Castner *et al.* deal with the Hall coefficient. Theories of the Hall coefficient are based on the three-site model originally developed by Holstein, described in Chapter 7. An adaptation to the case of variable range was given by Grunewald *et al.* (1981). The small Hall constant predicted has been observed by Koon and Castner (1987).

In InP Finlayson, Mason, and Mahomedd, and for n-InSb Mansfield *et al.* (1985) and Biskupski and Briggs (1988) have obtained similar results by compressing the wave function in a magnetic field. Biskupski *et al.* (1992) give further results, claiming that the transition between the low regime should occur when $\hbar/eaB \simeq d$, where d is the distance between impurities and a the Bohr radius in the absence of a field. A similar phenomenon is observed in GaS when the system is moved away from the neighbourhood of the transition by a magnetic field (Shlimak 1990).

In the hopping regime and for concentrations below it, the specific heat and magnetic properties, particular of doped semiconductors, are of great interest and have been intensively investigated. Since much of the book in this series by Kamimura and Aoki (1989) deals with this problem, we shall not deal with it here.

J. H. Davies *et al.* (1982) give reference to much experimental work in the Ioffe Institute in Leningrad (as it then was). Benzaquem and co-workers (1985) found 1/4 for the index in compensated GaAs and InP.

Various authors have found evidence for a Coulomb gap from data not related to hopping. Thus Franz and Davies (1986) found evidence in non-metallic sodium tungstate bronzes from the optical absorption coefficient. In a series of papers Whall *et al.* (1984, 1986, 1987) traced its influence on the conductivity and thermopower (see Mott 1991, p. 54).

In this section we have used elementary arguments to obtain the $T^{1/4}$ and $T^{1/2}$ laws. For a more accurate percolation method, which gives almost the same results, see Ambegaokar *et al.* (1971) and Pollak (1972). More recently Sivan *et al.* (1988) and others have argued that the percolative-like nature of the charge transport in these systems can give rise to a nonlinear averaging process that may cause a negative magnetoresistance. Effects on thin films are anticipated and have been investigated experimentally (Ovadyahu 1986; Erydman *et al.* 1992).

Aharoni *et al.* (1992) have shown that an equation can be found which describes a universal crossover from the $T^{1/2}$ to the $T^{1/4}$ behaviour. A problem that is unsolved (see Chapter 11) is the absence of a Coulomb gap for hopping in a two-dimensional impurity band (Timp *et al.* 1986).

Schirmacher (1991) gives a discussion of magnetoresistance through quantum interference in variable-range hopping; reference to earlier work is given in this paper.

Mansfield (1991) claims from a review of the experimental evidence that the $T^{1/4}$ law is normally obeyed except near the metal–insulator transition; this may be a similar conclusion to ours; only near the transition is $\langle E(T) \rangle$ not large compared with the Coulomb gap.

Finally, the problem of the excitations of a 'Fermi glass'—that is an electron gas in which states are localized at the Fermi energy, has not been finally resolved. The reader is referred to the paper on interacting particles by Mochena and Pollak (1991), and also to the book by Kamimura and Aoki (1989) and to Pollak (1992).

3.6. Conduction in granular metals

Another interesting case in which localization occurs is provided by conductivity in granular metal films. Abeles *et al.* (1975) and Ping Shen and Klafter (1983) investigated films prepared by co-sputtering metals (Ni, Pt, Au) and insulators (SiO$_2$ and Al$_2$O$_3$). Figure 3.7 shows the resistivity at 4.2 K and room temperature as a function of x, the volume fraction of Ni in the films.

It will be noticed that the transition from a positive to a negative temperature coefficient of resistance occurs when $\sigma \sim 10^2\,\Omega^{-1}\,\text{cm}^{-1}$, a reasonable value for σ_{min}; the distance between the particles being $\sim 50\,\text{Å}$. It appears that quantum interference does not occur in this case. The explanation could be that energy exchange between electrons in the metallic particles leads to inelastic scattering.

At the time of writing recent work is inconclusive about the proper model to apply to granular metals. Pollak and Adkins (1992) examine many models, and favour correlated hopping in a Coulomb gap.

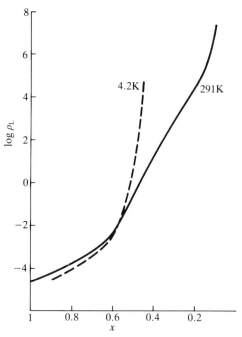

Fig. 3.7. Low-field resistivity ρ_L in Ω cm as a function of volume fraction x of Ni in Ni–SiO$_2$ sputtered films, at the temperatures shown (Abeles and Ping Shen 1974).

3.7. Weak localization

The Kawabata equation in the form

$$\sigma = \sigma_B \left\{ 1 - \frac{1}{(\kappa l)^2} \left(1 - \frac{l}{L_i} \right) \right\} \tag{3.28}$$

with L_i caused by electron–electron collisions, valid when the second term in the traces is small compared with unity, predicts a correction to σ/σ_B of the form const T (negative temperature coefficient of resistance) if L_i is caused by electron–electron collisions and thus proportional to $1/T$. As we shall see in Chapter 5, another term, due to Altshuler and Aronov, varies as $T^{1/2}$ and so should be predominent at low T, but often this term is not observed. In the literature the term in T is called 'weak localization' because it is caused by the term which when strong drives the system towards an Anderson transition. Several examples are given in this book. Figure 3.8 shows results from amorphous InO$_2$, the results being those of Graham *et al.* (1991).

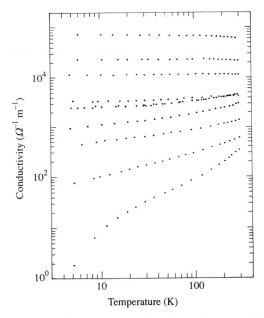

Fig. 3.8. Conductivity against temperature for the entire range of transparent films. High-conductivity films prepared at low oxygen pressure are at the top of the diagram, moving to lower-conductivity films towards the bottom as the oxygen pressure is increased.

3.8. Percolation transitions; metal–rare gas systems

Frozen mixtures of metals and rare gases have been investigated over a number of years, partly to determine the nature of the metal–insulator transition. Early work was by Cate *et al.* (1970) and Even and Jortner (1972), which is reviewed by Mott and Davis (1979, p. 156). Following more recent work by Micklitz and co-workers (Ludwig *et al.* 1981; Micklitz 1985) it appears that most of these mixtures have a granular structure; the conductivity at low temperatures near the transition follows that predicted by classical percolation theory. The conductivity near the transition behaves like $(x - x_c)^v$ where x is the concentration of metal and v lies in the range 1.5–2.0. The value predicted by classical percolation theory is 1.6 (Kirkpatrick 1973; see also § 5.3). The first system to show a random mixture on an atomic scale was Bi–Kr (Ludwig and Micklitz 1984). The evidence is that the super-conductivity transition temperature shows a strong decrease with decreasing concentration of metal, which is not the case for instance for Sn–Ar. For the former $v = 1.07 \pm 0.1$; the value expected is unity. The transition thus seems to be of the Anderson rather than the Mott (discontinuous) type (see Mott 1991, p. 211).

3.9. The scaling theory of Abrahams *et al*.

The important paper of Abrahams *et al.* (1979) first made clear that the conductivity $\sigma(E)$ of a degenerate electron gas must tend continuously to zero, as E_F tends to E_c. These authors defined a dimensionless conductance

$$G(L) = (Lh/e^2)\sigma(L)$$

where $\sigma(L)$ is the conductivity of a cube of side L. The proof depends on an argument of Thouless (1977) who wrote

$$G(L) \propto V(L)/W(L)$$

where $W(L)$ is the distance between quantized levels in the box, and $V(L)$ is the change in the energy levels resulting from a change in the boundary conditions, for instance from the vanishing of ψ to the vanishing of $\partial\psi/\partial n$ differentiated normal to the surface. For $V(L)$ we may estimate as follows. A change in boundary conditions will change the energy $\hbar^2 k^2/2m$ by

$$\hbar^2 k\delta k/m = \hbar^2 k/mL$$

if the mean free path l is infinite. However, for a finite l, only a thickness l around the boundary is affected, so

$$V(L) \propto (\hbar^2 k/mL)l/L.$$

Thus, writing $N(E) = mk/\hbar^2$, we find

$$G(L) = k^2 lL,$$

which is what we expect.

It is then argued that, if we fit together these blocks to form larger blocks, the only relevant quantity determining the new value of G is the old one, because it will determine the change of energy levels when hypercubes are fitted together. Thus, for a block made up of smaller ones of volume b^3, it is asserted that

$$G(b, L) = f[G(b)]$$

where f is some universal function. We may write this as a differential equation if we take an infinitesimal increase in scale size. Then we find

$$\mathrm{d}\ln G(L)/\mathrm{d}\ln L = \beta[G(L)]$$

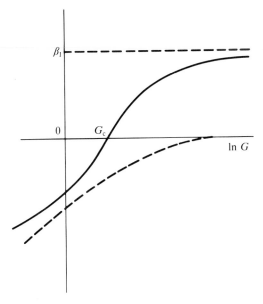

Fig. 3.9. β-function plotted against G (the dimensionless conductance) in three dimensions (full line) and in two dimensions (broken line).

where β is a universal function. For large G in three dimensions, perturbation theory shows that

$$\beta = 1 - a/G$$

and, for small G,

$$\beta = \ln(G/G_0).$$

Placing the two forms together we see that β must be as in Fig. 3.9. The important point is that β has a zero, so that in the neighbourhood of this zero

$$G(L) = \text{const}$$

which means that σ vanishes for $L \to \infty$. The zero, therefore, corresponds to the mobility edge and σ must vanish there.

The conductivity of a cube of side L will of course show fluctuations, and a proper averaging process is necessary (Anderson *et al.* 1980; Mott and Kaveh 1985*b*).

In two dimensions, perturbation theory shows that for large G

$$\beta = -a/G \tag{3.29}$$

so the β function is always negative as shown in Fig. 3.9. The deduction is that, in two dimensions, states are always localized. This conclusion is discussed in Chapter 10.

Asbel (1991) has shown that, in two-dimensional problems, while in the absence of a magnetic field any disorder will localize all states a field produces isolated localized states at the Landau energies.

3.10. Minimum metallic conductivity and the effect of magnetic fields

For many systems, for instance $La_{1-x}Sr_xVO_3$ (§ 6.7) and for impurity bands, the plot of resistance against temperature for varying composition was early shown to be as illustrated in Fig. 3.10, with the pre-exponential factor σ_0 of order σ_{min}. This is discussed by Mott (1989), and it is suggested that interaction with phonons (or possibly magnons) is strong enough to ensure that $L_i \sim a$ in these cases at relatively high temperatures, so that $\sigma_0 = 0.03e^2/\hbar L_i$. If so, we do not, therefore, expect a flat plot with $\sigma = \sigma_{min}$ as shown, and more detailed observations for $La_{1-x}Sr_xVO_3$ are needed.

Early work on doped indium antimonide (Biskupski *et al.* 1981; Manfield *et al.* 1988) suggested that a constant pre-exponential might exist down to low T. More recent work by Biskupski *et al.* (1992) shows, however, that this is not so; in § 5.1, where the effect of long-range interaction is taken into account, we show that for $\sigma < \sigma_{min}$, the temperature of the conductivity is

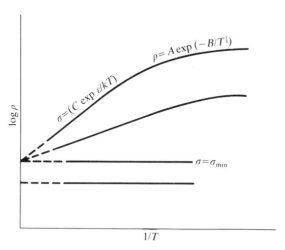

Fig. 3.10. Log ρ plotted against $1/T$ for varying values of the parameters if σ_{min} exists.

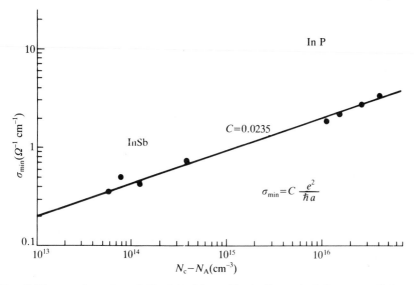

Fig. 3.11. $\sigma_{\min}$ for n-type InP plotted logarithmically against the mean distance between centres (Biskupski 1982).

given by

$$\sigma = \sigma_0 + AT^\nu$$

where σ_0 tends to zero as E tends to E_F and ν has the value $1/2$, tending to $1/3$ at the transition.

The extent to which the pre-exponential is often given by $\sigma_{\min}$ is shown by Fig. 3.11, where $\sigma_{\min}$ is deduced from curves of type (3.5), an activation energy being induced by a magnetic field. We believe that our explanation is compatible with a fairly sharp mobility edge, for the following reason. If we write (Mott 1985b)

$$\sigma_0 = e^2 N(E_c) D$$

where D is the diffusion coefficient, and take σ_0 as $0.03 e^2/\kappa L_i$ and $L_i = (D\tau_i)^{1/2}$, then the width of the mobility edge, defined by

$$\Delta E = \hbar/\tau_i$$

is easily seen to be given by

$$\Delta E/E_0 = 0.03 (a/L_i)^3 \tag{3.30}$$

where E_0 is defined by

$$N(E_c) = 1/a^3 E_0.$$

This is why even when L_i becomes comparable with a, the mobility edge can remain fairly sharp.

In the presence of a strong magnetic field, at the lowest temperatures a minimum metallic conductivity can, we believe, be observed. If we write

$$\sigma = \sigma_B g^2 \left\{ 1 - \frac{1}{(kag)^2} \left(1 - \frac{a}{L_H} \right) \right\}$$ (3.31)

the magnetic field will cause g to decrease because of shrinkage of orbits; L_H decreases, pushing σ up, but when $L_H = a$ the interference term vanishes and cannot decrease any further. If g at this point is greater than $1 < 3$, so that the material is metallic, σ will continue to decrease with decreasing g, until the value $g = 1/3$ is reached when $\sigma = \sigma_{min}$ and localization sets in at this point.

On the experimental side, Long and Pepper (1984) found evidence that this was so in InP. Biskupski *et al.* (1992) found on the contrary that σ, extrapolated to zero as $H - H_c$. The present author (Mott 1992) has suggested that the discrepancy arises because the amount of doping was not the same in both cases; in one the material is still metallic when $L_H = a$, in the other not. We do not think that this conclusion is affected by the inclusion of the interaction term (Chapter 5) in eqn (5.6), which at $T = 0$ just changes the constants slightly.

3.11. Thermopower

Measurements of the thermopower of non-crystalline semiconductors can show whether the carriers are electrons or holes; for amorphous metals, too, the same kind of information can be deduced as for crystalline materials, using eqn (3.34). To deduce the appropriate formulae, using a method first given by Cutler and Mott (1969), we start with the d.c. conductivity σ_E at zero temperature for electrons with energy E. This is the metallic conductivity at $T = 0$ when the Fermi energy is E. According to the Kubo–Greenwood formula, σ_E is given by eqn (3.5), but the corrections discussed in this chapter must be included and σ_E will tend to zero as $E \rightarrow E_c$.

Then for the conductivity σ we can write

$$\sigma = -\int \sigma_E (\partial f / \partial E) \, dE,$$ (3.32)

where f is the Fermi distribution function, and the thermopower S is then given by

$$S\sigma = \frac{k_B}{e} \int \sigma_E \frac{E - E_F}{k_B T} \frac{\partial f}{\partial E} \, dE.$$ (3.33)

The proof is as follows. If F is the field, then the current $\mathrm{d}j$ due to electrons with energies between E and $E\,\mathrm{d}E$ is given by

$$\mathrm{d}j = -\sigma_E \frac{\partial f}{\partial E} F\,\mathrm{d}E.$$

The free energy carried by this current is $-(E - E_F)\,\mathrm{d}j/e$, which becomes

$$\frac{1}{e}\frac{\partial f}{\partial E}\sigma_E(E - E_F)F\,\mathrm{d}E.$$

Integrating this expression we obtain the total electronic heat transport, which is equal to $j\Pi$, where Π is the Peltier coefficient, so that

$$\Pi j = \frac{F}{e}\int \sigma_E \frac{\partial f}{\partial E}(E - E_F)\,\mathrm{d}E.$$

Since $S = \Pi/T$, eqn (3.33) follows.

The following expressions can then be deduced. For metals the current and thermopower are determined by electrons with energies in the neighbourhood of E_F, so that the same expression as for crystalline metals follows, namely

$$S = \frac{\pi^2}{3}\frac{k_B^2 T}{e}\left\{\frac{\mathrm{d}\ln\sigma}{\mathrm{d}E}\right\}_{E=E_F} \tag{3.34}$$

For semiconductors in which a mean free path can be defined, we obtain the usual expression

$$S = \frac{k_B}{e}\left(\frac{E_c - E_F}{k_B T} + \frac{5}{2} + r\right)$$

where $r = \mathrm{d}\ln\tau/\mathrm{d}\ln E$ and τ is the relaxation time. For amorphous semiconductors, Cutler and Mott (1969) and Fritzsche (1971), assuming a discontinuity on σ_E at E_c, obtained

$$S = \frac{k_B}{e}\left(\frac{E_c - E_F}{k_B T} + A\right) \tag{3.35}$$

with $A = 1$. However, if, as we now believe, σ_E tends to zero as $E - E_c$, a value $A = 2$ would be appropriate. In Chapter 8 we will discuss which formulation is appropriate to the case of liquids, where, we shall argue, $l = L_i$.

At the time of writing, it is uncertain whether the quantity E_F in eqn (3.35) should be the true Fermi energy, which varies with temperature, or the value at zero temperature. Relevant papers are Emin (1977b, 1984), Butcher and Friedman (1977), Butcher (1984), Overhof and Beyer (1983), and Liu and Emin (1984).

If E_F lies above a mobility edge, so that the material is metallic, the behaviour of the thermopower S as E_F tends to E_c is of interest. We argue in Chapter 8 that a minimum metallic conductivity exists for liquids; if this is so, and $\sigma = \sigma_{min} + \alpha(E_F - E_c)$ just above the edge, then from (3.32)

$$S = \frac{\pi^2}{3} \frac{k_B^2 T}{e} \frac{\alpha}{\sigma_{min}},$$

giving a small value proportional to T. If, on the other hand, as in solids σ goes linearly to zero so that

$$\sigma = \alpha(E_F - E_c),$$

then eqn (3.34) gives

$$S = \frac{\pi^2}{3} \frac{k_B^2 T}{e} \frac{1}{E_F - E_c}, \tag{3.36}$$

which appears to diverge as $E_F - E_c$ tends to zero. However, an average over a range of energies of width $k_B T$ suggests that $1/(E_F - E_c)$ could be replaced by the integral

$$\int (E - E_c) \, dE$$

between the limits E_F, $E_F + k_B T$, giving

$$\ln\{(E_F + k_B T - E_c)/(E_F - E_c)\},$$

which implies that no divergence is expected and the thermopower should be of order $(\pi^2/3)k_B^2 T/e$. This could be considerably greater than the value given by (3.34).

At low temperatures, if the density of states at the Fermi energy is finite, charge transport will be by variable-range hopping. Treatments of the thermopower were given by Zvyagin (1973), Kosarev (1975), Overhof (1975), and Butcher (1976). Starting from (3.36), these authors integrated over the range of energies W which contribute to the thermopower, supposing this to be the activation for hopping, namely

$$W = k_B(T_0 T^3)^{1/4}. \tag{3.37}$$

Here T_0 is defined by the relation

$$\sigma \propto \exp(-T_0/T)^{1/4}.$$

We set for the conductivity

$$\sigma = e^2 N(E_F) D$$

where D is the diffusion coefficient. Then, supposing D to be independent of

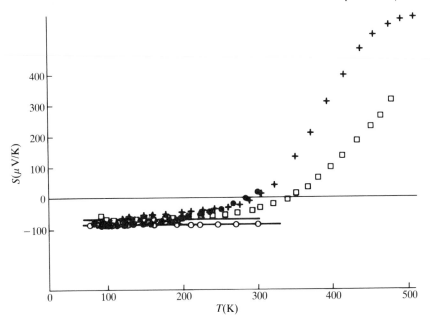

Fig. 3.12. Thermopower of sputtered and amorphous germanium, below 250 K probably in the hopping regime (Theye *et al.* 1985). (○) Sputtered at 25 °C; (●) evaporated at 25 °C; (+) evaporated at 25 °C and annealed at 350 °C, (□) evaporated at 220 °C. The theoretical value, $-(k_B/e)\ln 2$ is $-52\ \mu V\ K^{-1}$.

energy, (3.33) gives

$$S = \frac{1}{2}\frac{k_B}{e}\frac{W^2}{k_B T}\left(\frac{d\ln N(E)}{dE}\right)_{E=E_F}.$$

For variable-range hopping, from eqn (3.35)

$$W^2/k_B T = k_B(T_0 T)^{1/2}$$

so S varies as $T^{1/2}$ tending to zero as $T \to 0$. This behaviour, according to Movaghar and Schirmacher (1981), is to be expected only at low temperatures. For hopping to nearest neighbours, on the other hand, W is constant and S should diverge as $1/T$.

If the Coulomb gap is taken into account and $\sigma \propto \exp(-A/T^{1/2})$, Burns and Chaikin (1985) found that S tends to a finite value as T tends to zero.

These equations are essentially for spinless particles, and are valid only at temperatures at which a random orientation of spins contributes little to the entropy. In the limit of high temperatures there should be an additional term $(k_B/e) \ln 2$ in the thermopower. This is observed in some semiconductors in the hopping regime (Butcher 1976). Results of this kind for amorphous germanium (Theye *et al.* 1985) are shown in Fig. 3.12.

4 Heavily doped semiconductors

4.1. Uncompensated semiconductors; the Mott and Anderson transitions

In earlier chapters we have discussed the behaviour of impurity bands in semiconductors as examples of systems, with properties close to those described by Anderson's (1958) model. In this chapter we examine such systems in greater detail.

In a doped semiconductor, in the absence of compensation, for low values N_D of the concentration of donors, the activation energy ε for conduction is finite and the conductivity tends to zero with temperature. As the concentration increases, ε decreases and eventually vanishes. Conductivity is then 'metallic'. The transition from non-metallic to metallic behaviour has sometimes been called the 'Mott transition' (Mott 1949). Two facts are important in understanding this behaviour.

1. The random positions of the donor centres.
2. The intra-atomic interaction of a pair of electrons within a centre.

First we neglect the former. We describe the system by a 'tight binding approximation' (§ 3.2). If there were no interaction between electrons, a band of levels would be formed as in Fig. 4.1, and this band would be half full. The material would then always show metallic conductivity, though for small values of N_D the effective mass would be large. This is certainly not the case; the conductivity can tend to zero as N_D falls.

The key to our understanding of this is the intra-atomic interaction between two electrons, measured by the 'Hubbard U', defined by

$$U = \int \int \frac{e^2}{\kappa r_{12}} |\psi(x_1)|^2 |\psi(x_2)|^2 \, d^3x_1 \, d^3x_2 \qquad (4.1)$$

where κ is the background dielectric constant and ψ the wave function for one of the centres. Intuitively one can understand that interaction will prevent an electron from moving into another centre which is already occupied. To treat this concept quantitatively, we introduce the two 'Hubbard bands'. Suppose an electron is removed from one of the centres, and placed on another at some distance from it. Then the 'hole' can move with a definite

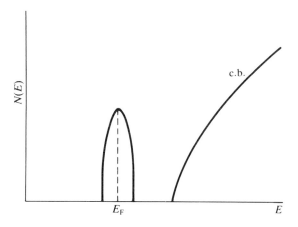

Fig. 4.1. Density of states $N(E)$ in a doped semiconductor with an impurity band, when interaction between electrons is neglected. E_F is the Fermi energy, and c.b. stands for conduction band.

wavenumber **k**, giving rise to a band of width B_1. This is called the 'lower Hubbard band'. In the same way, the electron placed on a site already occupied can move, and it too produces a somewhat broader band of width B_2. The Hubbard U is equal to the separation between the centres of the two bands. Then the system is an insulator if

$$U > \tfrac{1}{2}(B_1 + B_2).$$

If, however, the concentration of donors should increase B_1 and B_2 will increase too and then metallic behaviour will begin when the bands overlap, which will occur when

$$U = \tfrac{1}{2}(B_1 + B_2). \tag{4.2}$$

At first sight it looks as if, as the concentration increases, the number of carriers should then increase continuously from zero. But this is not so. Consider any two bands, either of Hubbard type or those formed by the crystalline structure, with a small energy gap ΔE between them. We illustrate two such bands in Fig. 4.2. Then it is argued that, as ΔE is decreased, at zero temperature there is a sudden increase in the concentration n of carriers from zero to a finite value (Brinkman and Rice 1973). The reason is that the energy of the system is of the form

$$\frac{3}{5}\frac{h^2 n^{2/3}}{m} - \frac{Ce^2 n^{1/3}}{\kappa} + \Delta E.$$

The first term, with m a reduced mass, represents the kinetic energy of the

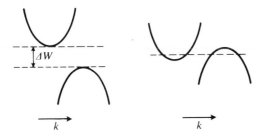

Fig. 4.2. Two bands in an indirect semiconductor, before and after a metal–insulator transition.

carriers, the second, with κ a dielectric constant, the interaction between electrons and holes. To determine the constant C, careful consideration of correlation is necessary (see Rice 1977). The important point is, however, that for a value of n given by

$$n^{1/3} = \text{const } me^2/\hbar^2\kappa, \tag{4.3}$$

the first two terms in the energy have a minimum *negative* value

$$E_0 = -\text{const } me^4/\hbar^2\kappa^2 \tag{4.4}$$

so that, when $\Delta E + E_0$ reaches zero, there is a discontinuous jump in the concentration n from zero to the value given by (4.3). We believe the discontinuity to be greater for Mott transitions than for band-crossing transitions, because for the latter κ can be very large near the transition, while for Mott transitions the virtual transitions which determine κ involve displacement of the electron from one centre to another. These transitions are likely to have a small oscillator strength, so that κ will be small. The value of n given by (4.3) will then be large (Mott 1974).

In doped semiconductors there is no evidence of a discontinuous transition. This is a result of the large measure of disorder which results from the random positions of the donors (cf. Mott 1977). There is, however, indirect evidence of a discontinuity in certain liquids such as caesium at high temperatures and decreasing densities, and in metal–ammonia solutions as a function of the concentration of metal; these are described in Chapter 8.

The concentration of donors at which the transition takes place has been predicted by various methods to be given by

$$n^{1/3}a_{\text{H}} \simeq 0.25. \tag{4.5}$$

Figure 4.3, taken from Edwards and Sienko (1978), shows how well this equation is obeyed for a wide variety of materials. For theoretical derivations, see for instance Mott and Davies (1980) and Mott (1991).

In principle, one has to equate U, given by (4.1), to $\frac{1}{2}(B_1 + B_2)$. Sometimes

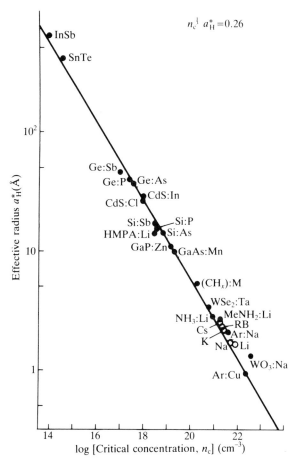

Fig. 4.3. Plot of effective radius in the equation $n_c^{1/3} a_H = 0.26$ against the logarithm of n_c (Edwards and Sienko 1978).

the difference between B_1 and B_2 has been neglected, so that one can write, with z the number of nearest neighbours,

$$B = 2zI$$

and, as shown in § 3.2,

$$I = I_0 \exp(-a/a_H).$$

With $1/a = n^{1/3}$, the transition is expected when

$$n^{1/3} a_H = 1/\ln(2zI_0/U)$$

with

$$I_0 = e^2\alpha(1 + \alpha R) = 5e\alpha$$

and

$$U = 5e^2\alpha/s$$

(Schiff 1955). The transition occurs when $n = n_c$ and

$$n_c^{1/3}a_H = 1/\ln(16z); \tag{4.6}$$

with $z = 6$ the right-hand side is 0.23. Since the value depends on the logarithm of a large number, improvements in the theory will affect the result only slightly.

As shown, for instance, by Newman and Holcomb (1983), experiment shows considerable difference between the experimental values of (4.6); thus n_c is some 40 per cent higher for Si:As then for Si:P.

For the application of these principles to the insulating state of materials such as NiO, see a series of papers by Baird (1977, 1990).

We have said above that there is no evidence for a discontinuity in the metal–insulator transition in uncompensated semiconductors, and suggested that its absence may be caused by the random positions of the centres, as described by Mott (1978c). If so, though the concentration at which the transition occurs will be determined approximately by the Hubbard U as described above, it will be observed when the lower and upper Hubbard bands overlap sufficiently for states at the Fermi energy to be delocalized.

Figure 4.4 shows what may happen when the two bands, broadened by

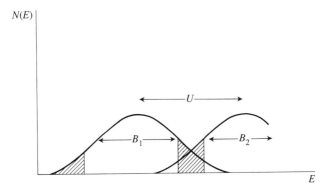

Fig. 4.4. Schematic illustration of how two Hubbard bands, with localized tails (shaded), resulting from disorder, can overlap, so that the equation $\frac{1}{2}(B_1 + B_2) \approx U$ determines approximately the concentration at which the transition occurs, while the properties of the materials near the transition are those resulting from a transition of Anderson type.

disorder, overlap. A region of Anderson-localized states may occur between E_c and E_c', which will disappear as the overlap increases. An example where this certainly occurs is a-Si$_{1-x}$Au$_x$, where Nishida *et al.* (1983) recorded a metal–insulator transition at $x = 0.14$, of Anderson type because variable-range hopping is observed on the insulating side, but a correlation gap occurs as the transition is approached on the metallic side. The authors found $v = 1$, which they consider consistent with strong spin–orbit interaction. The transition is thus essentially of Anderson type, with no discontinuity and σ tending continuously to zero. The discussion given in the next section suggests that for materials such as Si:P, this is indeed the case.

Another way of treating the Mott transition, which goes back to Herzfeld (1927), is through the use of the Clausius–Mossotti relationship, which relates the dielectric constant κ to the atomic polarizability α by the relationship

$$\kappa = 1 + \frac{4\pi N\alpha}{1 - (4\pi/3)N\alpha}.$$

If the denominator vanishes, a transition to the metallic state should occur. Various authors (Berggren 1974; Edwards and Sienko 1983) have shown that this equation leads to satisfactory numerical results. In my view, as the dielectric constant increases, so does the radius $\hbar^2\kappa/me^2$, and the overlapping of the wave functions leads to the breakdown of the above formula (for a description of this overlap see Mott and Gurney 1940, Chapter 1). If a discontinuous transition is predicted, it will occur before κ diverges.

That κ diverges at an Anderson transition is shown for instance by Mott and Kaveh (1985b).

4.2. Impurity conduction; the observed behaviour

The idea of impurity conduction in a doped and compensated semiconductor was first introduced by Gudden and Schottky (1935), who mentioned the possibility of a process in which a carrier is transferred from an occupied to an equivalent empty donor of higher energy. This is the hopping process already described in § 3.5. It is immediately clear from the considerations of the last section that this process can only take place in a *compensated* material. Busch and Labhart 1946) were the first to observe it, in SiC, and Hung and Gleissman (1950) in germanium. The conductivity shows a small activation energy, ε_3, depending on concentration and compensation, and a small pre-exponential factor varying as $\exp(-2R/a_H)$, where R is the mean distance between centres and a_H the hydrogen radius. The treatment given by Miller and Abrahams (1960) of hopping to nearest neighbours has already been described (§ 3.5), as has the process of variable-range hopping expected at low temperatures (Mott 1968).

Our model then is of a tight binding lower Hubbard band, not fully occupied, in which Anderson localization is produced by the random field of the charged acceptors and by the random positions of the donors. As the concentration is increased, the bandwidth B should increase and eventually states in mid-band will become delocalized. A mobility edge will then exist between localized and delocalized states. As the concentration increases, the mobility edge will eventually move to the Fermi energy, whereupon the activation energy for conduction (denoted by ε_2) will disappear and a transition to metallic conduction will take place. I have called this an

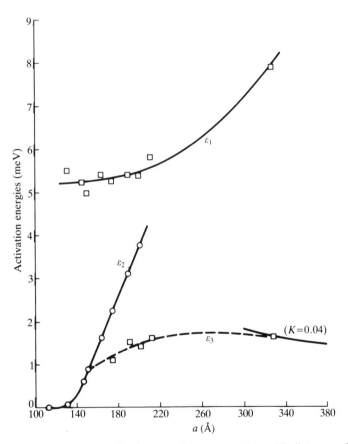

Fig. 4.5. Variation of the activation energies ε_1, ε_2, and ε_3 with distance a between donors for n-type germanium (Davis and Compton 1965). For ε_3 the calculations of Miller and Abrahams for $K = 0.04$ are shown (full curve).

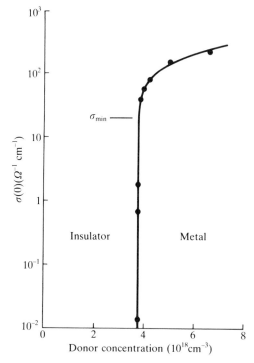

Fig. 4.6. Conductivity of Si:P extrapolated to zero temperature as a function of donor concentration (Rosenbaum *et al.* 1980).

'Anderson transition'. It is quite different from the 'Mott transition' described in the last section; it depends essentially on disorder, and the Hubbard U (intra-atomic interaction) plays no essential role. Also no discontinuity in n is expected.

The way the activation energies behave in doped germanium is shown in Fig. 4.5 from Davis and Compton (1965). ε_1 is the energy required to excite an electron into the conduction band, ε_2 that described above for excitation into the upper Hubbard band (called by Fritzsche the D^- band), and ε_3 the hopping energy. It will be seen that the conductivity σ goes continuously to zero as the concentration is decreased.

The compensation in this work was quite low (4 per cent). As suggested in the last section, we believe that for both compensated and uncompensated materials, the transition is of the Anderson type. Bhatt and Rice (1981) maintained, as suggested by Fritzsche (1978), that this was so for many-valley conduction bands, as in n-type silicon doped, for instance, with phosphorus. If so, it is remarkable that the transition point appears to satisfy an equation derived from the assumption that the transition is of the Mott–Hubbard type. The reason is that the conditions for the two types of transition are similar. The Mott–Hubbard transition takes place (eqn (4.2)) when $B = U$ leading to $n^{1/3}a_H = 1/\ln 96$. The Anderson transition (Mott and Kaveh 1985b, p. 375) should occur when $V_0 = 1.7\,B$, where V_0 is the random potential. For B we write

$$B = 2z(e^2\alpha/\kappa)(1 + \alpha R)\exp(-\alpha R)$$

where $\alpha = 1/\xi$ and a is the distance between centres. B is of the order e^2/ξ. Since αa is of order 4 and $z = 6$, this leads to

$$\alpha a \sim \ln 100 = 4.6$$

and with $\alpha a = 1/n^{1/3}a_H$, this is very close to the condition for the Mott transition.

All the evidence shows that, in the absence of a magnetic field, at zero temperature σ tends continuously to zero, as it does in other similar systems.

Examples of this behaviour are found in the work of Hertel $et\ al.$ (1983) on amorphous Nb–Si, where σ at the lowest temperatures tends linearly to zero with decreasing content of niobium (Fig. 4.5), and the work of Thomas and co-workers (Rosenbaum $et\ al.$ 1980; Thomas 1983) on silicon doped with phosphorus, the conductivity being measured down to a few millikelvin and extrapolated to zero temperature (Fig. 4.6). Here the conductivity appears to vary as $(n - n_c)^{1/2}$. The reason for this is described in § 5.4.

5 Effect of long-range interaction between electrons on the electrical properties of a Fermi glass

5.1. The Altshuler–Aronov correction

Altshuler and Aronov (1979) first predicted an important effect of the long-range interaction between electrons in a condensed gas in a disordered medium, the most striking results being the addition of a term $mT^{1/2}$ to the conductivity, so that

$$\sigma = \sigma_0 + mT^{1/2} \tag{5.1}$$

m can have either sign. We shall discuss the effects of this term using the hypothesis of Kaveh and Mott, that we can represent it by adding a term to the modified Kawabata equation (3.12), which, if necessary, we extrapolate to the transition. This term applies to the metallic range for either compensated or uncompensated material. It applies to all cases where the mean free path l is short, whether the degenerate electron gas is in a conduction band or in an impurity band. The extra term in the resistivity comes from a change in the density of states $N(E)$ near the Fermi energy in a metal, as illustrated in Fig. 5.1. The correction may be either positive or negative.

Before considering this term we summarize the other effects of electron–electron interaction on the resistivity. These are:

1. The well-known T^2 term first predicted by Landau and Pomeranchuk (1936) and by Baber (1937) resulting from collisions between electrons.

2. A term in eqn (3.12) resulting from the variation of L_i with temperature, given by

$$\delta\sigma/\sigma = C/L_i(k_F l)^2 \tag{5.2}$$

If L_i is the result of a collision of the electron considered as carrying the current with another such electron (the Landau–Baber process), we can write (see § 3.2)

$$L_i = (D\tau_i)^{1/2},$$

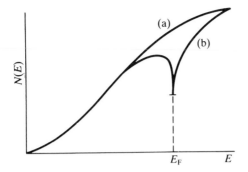

Fig. 5.1. Change in the density of states resulting from electron–electron interaction.

with τ_i, the time between inelastic collisions, given by

$$\hbar/\tau_i \sim (k_B T)^2/E_F.$$

We then find that

$$\frac{1}{L_i} \sim \frac{k_B T}{\sqrt{(v_F l E_F \hbar)}}, \qquad (5.3)$$

giving an *increase* in conductivity with τ linear in T, and again significant in the case of small l. In doped silicon at very low temperatures both terms have been observed, as they have also at much high temperatures in amorphous metals (see § 5.5; Howson 1984).

If L_i is the result of collisions with phonons, $1/\tau_i$ is proportional to T and, as we have seen in § 3.3, $\Delta\rho/\rho$ will vary as $T^{1/2}$.

As we stated at the beginning of this chapter. Altshuler and Aronov (1979) showed that the change in the density of states $\delta N(E)$ resulting from electron–electron interaction is of the form of Fig. 5.1. It is given by

$$\delta N(E) = \frac{C}{2\pi^3 \hbar D l} \left\{ -1 + l \left(\frac{E - E_F}{\hbar D} \right)^{1/2} \right\}. \qquad (5.4)$$

We give a proof in § 5.3. Starting from eqn (5.4), however, and supposing that $l \sim a$ (the Ioffe–Regel regime), σ is proportional to g^2, so that $\delta\sigma/\sigma = 2\delta N/N$, where putting $|E - E_F| = kT$ we find

$$\frac{\delta\sigma}{\sigma} = \frac{C}{\pi^3 \hbar D l N(E_F)} \left(1 - \frac{l}{L_T} \right). \qquad (5.5)$$

Here L_T, which we call the interaction length, is given by

$$L_T = (\hbar D/k_B T)^{1/2}.$$

The Kawabata equation (3.12) then becomes

$$\sigma = \sigma_B g^2 \left\{ 1 - \frac{1}{(k_F l g)^2} \left(1 - \frac{l}{L_i} \right) - \frac{C}{(k_F l g)^2} \left(1 - \frac{l}{L_T} \right) \right\} \qquad (5.6)$$

where, if the interaction is of the exchange type,

$$C = 1/\pi \hbar D l N(E).$$

We have, however, to consider the contribution to C from direct electrostatic interaction, the so-called Hartree term. Finkelstein (1983) and Altshuler and Aronov (1983) argued that, when the Hartree term is included, C must be multiplied by

where

$$\left.\begin{array}{c} \tfrac{4}{3} - \tfrac{2}{3}\tilde{F} \\[8pt] \tilde{F} = 32(1 + \tfrac{1}{2}F)^{3/2} - (1 + \tfrac{3}{4}F)/3F \\[8pt] F = x^{-1} \ln(1 + x) \end{array}\right\} \qquad (5.7)$$

and

where $x = (2k_F \lambda_s)^2$, λ_s being a screening constant. If $\tilde{F}$ is large enough the value of C changes sign. Kaveh and Mott (1987) argue that this normally occurs for n-type many-valley semiconductors, which accounts for the abnormal sign of m in these materials, and (see § 3.8) another abnormal index in a metal–rare gas mixture (cf. Mott 1991, p. 153).

5.2. The $T^{1/3}$ behaviour near the transition

In the case where C is positive, at the transition the conductivity should vary as $T^{1/3}$, and near the transition $\sigma = a + mT^{1/3}$ is observed. An example due to Maliepaard et al. (1988) is shown in Fig. 5.2.† The theoretical explanation, valid only for positive C in (5.6), is as follows. Near the transition, in the term

$$\frac{C}{(k_F a g)^2} \frac{a}{L},$$

which will determine the conductivity, L_T is equal to $(\hbar D/k_B T)^{1/2}$, but D is itself proportional to σ through the Einstein equation

$$\sigma = e^2 D N(E_F). \qquad (5.8)$$

† Earlier results by Newson and Pepper (1986) for GaAs and InSb (n-type) shows similar behaviour.

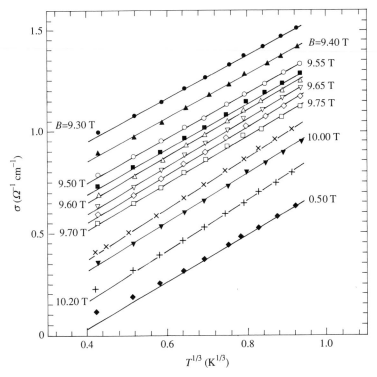

Fig. 5.2. Conductivity of doped and compensated GaS in various magnetic fields near the transition, plotted against $T^{1/3}$ (Maliepaard *et al.* 1988).

Thus we have

$$\sigma = \frac{e^2}{3\hbar a} \frac{C}{(k_F a)^2} a \left(\frac{k_B T N(E_F) e^2}{\sigma} \right)^{1/2}, \tag{5.9}$$

whence it is easily seen that $\sigma^{3/2}$ is proportional to $T^{1/2}$, so $\sigma \propto T^{1/3}$.

An interesting feature of Fig. 5.2 is that some specimens show metallic behaviour only above a certain temperature, and σ is then of the form $a + bT^{1/3}$, with a negative value of a. Below this temperature they are thought to be insulators, with conduction by hopping. We believe, following Thouless (1977), that when L_i or L_T become smaller than ξ localization can no longer occur and conduction will show metallic behaviour.

5.3. Deduction of the Altshuler–Aronov correction term in the density of states

In this section we deduce eqn (5.4) for the correction to $N(E)$ resulting from long-range interaction, first predicted by Altshuler and Aronov (1979). The derivation is similar to that given by Mott and Kaveh (1985a) and Mott (1991). We have to consider the interaction $V(|\mathbf{r}_1 - \mathbf{r}_2|)$ between a pair of electrons. Taking only the exchange term, the resulting self-energy $S(E)$ for a given electron with energy E is

$$S(E) = -\sum_{E'} \int\int d^3r_1\, d^3r_2\, V(|\mathbf{r}_1 - \mathbf{r}_2|) \psi_E^*(\mathbf{r}_1)\psi_E^*(\mathbf{r}_2)\psi_{E'}(\mathbf{r}_1)\psi_{E'}(\mathbf{r}_2).$$

We write

$$V(|\mathbf{r}_1 - \mathbf{r}_2|) = \frac{1}{(2\pi)^3} \int d^3q\, V(q)\exp\{i\mathbf{q}\cdot(\mathbf{r}_1 - \mathbf{r}_2)\}$$

which defines $V(q)$. We then see that

$$S(E) = -\frac{1}{(2\pi)^3} \int dE'\, N(E') \int_{ql<1} d^3q\, V(q)|\langle\psi_E^*|\exp(i\mathbf{q}\cdot\mathbf{r})|\psi_{E'}\rangle|^2.$$

A short calculation (see Mott 1991, p. 88) gives

$$S(E) = \frac{1}{8\pi^4\hbar N(E_F)} \int dE'\, N(E') \int_{ql<1} d^3q\, V(q)\, \frac{Dq^2}{(Dq^2)^2 + (E - E')/\hbar^2}.$$

If the interaction $V(r)$ is screened according to the Thomas–Fermi law $V(r) = (e^2/\kappa r)\, e^{-r/\lambda}$ then

$$V(q) = \frac{1}{N(E_F)a^3},$$

the charge e cancelling out because $-\lambda^2 = \kappa/e^2 N(E_F)$, Thus the change in the density of states, given by

$$\delta N(E) = \frac{\partial S(E)}{\partial E}, \quad E = E_F,$$

becomes

$$\delta N(E) = \frac{1}{2\pi^3\hbar Dl}\left\{-1 + l\left(\frac{|E - E_F|}{\hbar D}\right)^{1/2}\right\}. \tag{5.10}$$

This is (5.4), which is what we set out to prove in this section.

5.4. The values $v = 1$ and $v = \frac{1}{2}$ of the index

As we have seen, when, at $T = 0$, x in a material such as $Si_{1-x}:P_x$ or a-$Si_{1-x}Nb_x$ approaches the composition for the metal–insulator transition of the Anderson type, the conductivity σ tends to zero as

$$\sigma = \sigma_0 \left\{ (n - n_c)/n_c \right\}^v$$

where v approximates to unity in many materials (compensated semi-conductors, a-Si_{1-x}–Nb_x for example), while others to be discussed in this section show $v = \frac{1}{2}$. These are the many-valley semiconductors for which m is positive in the equation $\sigma = \sigma_0 + mT^{1/2}$. For this case, formulae given above suggest that

$$\sigma = \sigma_0 g^2 \left(1 + \frac{C}{2\pi^3 N(E_F)\kappa Da} \frac{1}{(gk_F a)^2} \right)$$

with C positive.

Since by Einstein's equation

$$\sigma = N(E_F)e^2 D$$

this reduces to

$$\sigma = \sigma_0 g^2 \{ 1 + Ce^2/2\pi^3 \hbar a\sigma (gk_F a)^2 \}.$$

Near the transition where $\sigma \to 0$ the second term is much greater than 1, so

$$\sigma^2 = C\sigma_0 e^2/2\pi^3 \hbar (k_F a)^2.$$

σ_0 should be the value of σ without the Altshuler–Aronov term, so $\sigma_0 = 0.03e^2/\hbar\xi$. Hence,

$$\sigma^2 = 0.03Ce^2/2\pi^2 \hbar^2 a\xi.$$

C is a number deduced from the Hartree interaction. Assuming that $C = 1$, we find

$$\sigma^2 \simeq \frac{3}{2}\sigma_{min} \frac{e^2}{\hbar\xi}.$$

Since ξ should vary linearly with n in a disordered system, we see that this model is able to account for the observed variation of the conductivity as $(x - x_0)^{1/2}$.

Kaveh and Mott (1992) have discussed some results of Micklitz and co-workers on deposited films of metals with rare gas. Only Ga–Ar and Bi–Kr are thought to give Anderson transitions (Mott, 1991, p. 209), the others having granular structure and $v = 1.6$, as for a percolation transition (§ 3.8). Ludwig and Micklitz (1984) found only for Bi–Kr that the super-conducting transition temperature T_c drops to zero continuously as the composition approaches that for the metal–insulator transition, while in all the others (e.g. Sn–Ar) T_c remains constant until the conductivity disappears. Of the materials mentioned above, the first has $v = \frac{1}{2}$ and the second $v = 1$. They are both superconductors, and for such materials above T_c, one expects the attraction between electrons to be greatly enhanced, then increasing the Coulomb term, which can account for the value $v = \frac{1}{2}$. In the bismuth compound, the very strong spin–orbit coupling can have a major effect on the sign of m in the equation $\sigma = \sigma_0 + mT^{1/2}$ in many-valley conduction bands, because it removes the valley degeneracy, on which the positive (Hartree) term depends, and thus favours $v = 1$. But this is not so for the bismuth–rare gas films, so it does not affect the index.

5.5. Effects of interaction in amorphous metals

We have seen in Chapter 2 that the temperature coefficient of resistance of amorphous metals can have either sign, and that a finite value of $\rho^{-1}\,\mathrm{d}\rho/\mathrm{d}T$ can persist down to very low temperatures. This cannot be explained by any model which depends on interaction with phonons. Kaveh and Mott (1983) suggested that the effects described in this chapter could be responsible, and a full experimental investigation was made by Howson (1984) and by Cochrane and Strom-Olsen (1984). A review of the subject is given by Howson and Gallager (1988). Phenomena which in doped semiconductors can be observed only below ~ 1 K are apparent in these materials at much higher temperatures.

At the lowest temperatures, we expect the term (5.4) varying as $T^{1/2}$ to predominate in the temperature coefficient of resistance of all amorphous metals. Owing to the term $D^{-3/2}$ in this equation, the term will be large for materials of high residual resistance. As the temperature rises, the linear term, arising from the relation $L_i \sim 1/T$ as a consequence of electron–electron collisions, will come into play. Above the Debye temperature, L_i is determined by collisions with phonons so that, since $L_i \propto 1/T$, $\delta\sigma/\sigma$ should vary as $T^{1/2}$. All these effects have been observed.

Figure 5.3 shows $\Delta\sigma$ plotted on a log:log scale against temperature for $Cu_{50}Hf_{50}$ and $Cu_{50}Zr_{50}$ (Howson and Grieg 1983). For both materials σ is of order $5 \times 10^3\ \Omega^{-1}\,\mathrm{cm}^{-1}$ ($5 \times 10^5\ \Omega^{-1}\,\mathrm{m}^{-1}$), so the correction is of order 1 per cent. The term in $T^{1/2}$ predominates below 25 K, the linear term above. Figure 5.4 shows similar results for $Cu_{50}Ti_{50}$ and $Ti_{50}Be_{40}Zr_{10}$. The absence of the $T^{1/2}$ term for the latter might be due to a small value of the factor $\frac{4}{3} - 2F$. Figure 5.5 shows some high-temperature results from Howson (1984)

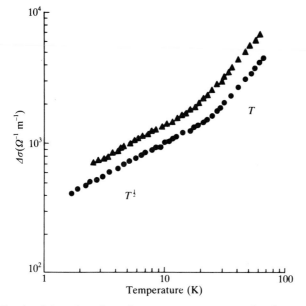

Fig. 5.3. Conductivity plotted against temperature on a log:log scale for (●) $Cu_{50}Ti_{50}$ and (▲) $Cu_{50}Zr$.

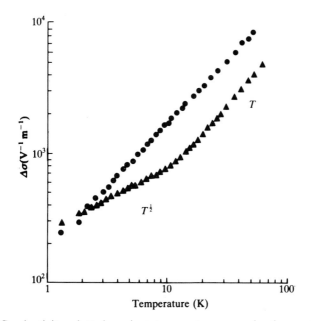

Fig. 5.4. Conductivity plotted against temperature on a log:log scale for (▲) $Cu_{50}Ti_{50}$ and (●) $Ti_{50}Be_{40}Zr_{10}$.

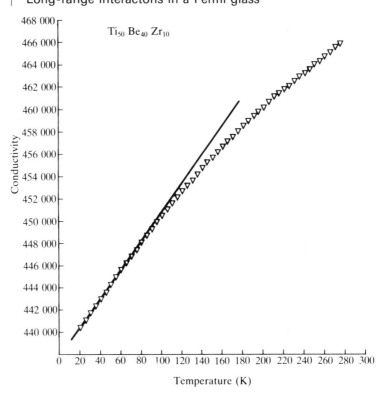

Fig. 5.5. Conductivity in Ω^{-1} cm^{-1} versus temperature for $\text{Ti}_{50}\text{Be}_{40}\text{Zr}_{10}$.

for $\text{Tl}_{50}\text{Be}_{40}\text{Zr}_{10}$, showing how $T^{1/2}$ behaviour sets in as the Debye temperature (~ 326 K) is approached.

In the correcting term in eqn (3.9) in a magnetic field H, L should be the cyclotron radius

$$L_H = (C\hbar/eH)^{1/2},$$

so a negative magnetoresistance varying as $H^{1/2}$, if observed, is the true test of the correctness of this theory of the temperature dependence. One expects the linear behaviour only if $L_H > L_i$; for lower fields a formula is given by Altshuler and Aronov (1979). Measurements on amorphous $\text{Cu}_{57}\text{Zn}_{43}$ were made by Bieri *et al.* (1984) and are shown by the dotted curves in Fig. 5.6. The full curves show the theoretical results. The deviations at high fields are accounted for in terms of spin–orbit interactions.

These magnetic field effects were not found by Lüescher *et al.* (1983) on similar amorphous metals under pressure; an interpretation in terms of interaction with phonons was given by Belitz and Schirmacher (1983).

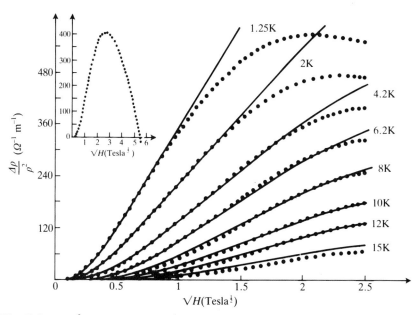

Fig. 5.6. $\Delta\rho/\rho^2$ olotted against $\sqrt{H}$ for amorphous $Cu_{57}Zr_{43}$ (dotted lines). The full lines refer to calculatons. The temperatures are shown. The inset shows data with pulsed fields. (From Bieri *et al.* 1984.).

6 Polarons

6.1. Introduction

An electron in a localized state, in either a crystalline or a non-crystalline material, will always distort its surroundings to some extent. For shallow donors in silicon and germanium, and for localized states in band tails, the radius of the localized state is normally large and the distortion consequently small; it is usually neglected, though this is not the case for deep levels. However, even for a free electron in a conduction band (or a hole in a valence band), significant distortion of the surroundings can occur under certain conditions, such as large effective mass. If distortion occurs, a 'pseudoparticle' is formed, which can move as a whole, the electron and the distortion it produces moving together. This pseudoparticle is called a polaron.

In the literature the first proposal for what we now call a polaron was in the paper by Landau (1933), discussed later in this chapter. Landau showed that in an *ionic* lattice, self-trapping could always occur. The next advance was the paper by Fröhlich *et al.* (1950), also for ionic solids with the medium treated as a polarizable dielectric without structure which leads to the theory of the 'large' or Fröhlich polaron, highly mobile and with relatively small bonding energy. It is the 'small' polaron that will concern us here, where the radial extension is comparable with the lattice constant and the mass enhancement larger. At low temperatures it behaves like a particle with enhanced effective mass m_{eff} and a mean free path l tending to infinity as the temperature tends to zero. The polaron is strongly scattered by phonons; at a temperature near $\frac{1}{2}\Theta_D$, where Θ_D is the Debye temperature, l becomes comparable with the distance between possible sites and at higher temperatures thermally activated hopping sets in, the mobility μ obeying the relation

$$\mu \propto \exp(-W_H/k_B T) \tag{6.1}$$

where W_H is the hopping energy. Polarons are not formed for electrons or holes in, for example, crystalline silicon, but are formed for holes in valence bands of alkali halides and solid rare gases, when they are called V_K centres. Though in principle mobile, they will be strongly scattered, trapped, or Anderson localized by impurities or disorder. The theory was first given by Holstein (1959) and will be presented in § 6.2.

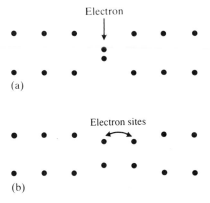

Fig. 6.1. (a) A molecular polaron and (b) the excited state that must be formed before an electron can hop from one site to another.

6.2. The Holstein polaron

We consider the idealized case of a hole in a simple cubic lattice of rare gas atoms, illustrated in Fig. 6.1. We consider the hole as being shared between two of the atoms; they attract and move towards each other, thus lowering the energy. We plot in Fig. 6.2 the energy of the system against q, the displacement. Aq^2 is the elastic energy. The displacement produces for the hole a 'potential well' of radius about a, the distance between atoms, and depth $H = Bq$, where B is a constant. It is a well-known result of quantum mechanics that a bound state in such a well exists only if

$$m_{\text{eff}} H a^2 / \hbar^2 > 1,$$

so the well will not trap a carrier or lower its energy until q exceeds a critical value q_0; the energy of the system without the term Bq_0 will thus appear as in curve (b) of Fig. 6.2. We write the distortion energy as $-B(q - q_0)$ (though the energy behaves like $(q - q_0)^2$ near q_0). Then the total energy E is given by

$$E = Aq^2 - B(q - q_0), \tag{6.2}$$

which has a minimum value E_0 when $q = B/2A$, given by

$$E_0 = Bq_0 - \tfrac{1}{4}B^2/A. \tag{6.3}$$

The energy E will necessarily have this minimum, but its value at this point can be positive or negative; only in the latter case will a stable self-trapped particle (i.e. a small polaron) form. This is most likely to occur for large effective mass, and thus for holes in a narrow valence band or for carriers in d-bands. If the polaron is unstable, there is practically no change in the

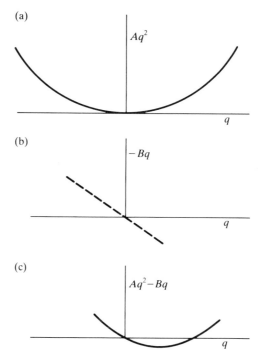

Fig. 6.2. Terms in the energy of an electron as a function of a configurational parameter q.

effective mass of an electron or hole in equilibrium in the conduction or valence band.

It will be seen that a barrier exists resisting self-trapping; this has been observed as a time delay by Laredo *et al.* (1981, 1983) for holes in AgCl, indicating a barrier height of 1.8 meV (cf. § 6.5).

The carrier can move either by excitation out of this self-trapped state into the conduction band or by 'hopping' to a neighbouring site. For hopping to occur, the interatomic distances between two adjacent pairs must be equal as illustrated in Fig. 6.1(b). Then the electron can move freely from one such pair to the other. In some cases the electron can move backwards and forwards several times before the system relaxes; the transition is then called adiabatic. Alternatively the chance of transfer in the time during which the configuration persists may be small.

At high temperatures the configuration of Fig. 6.1(b) is produced by thermal excitation, and the motion is by 'thermally activated hopping', with mobility proportional to a factor of the type (6.1). To find W_H, we must ask what displacement of the kind shown in Fig. 6.1(b) has the smallest energy.

This will be so when both displacements are equal to $B/4A$, as may easily be verified. In eqn (6.3) E_0 then takes the form

$$W_p = \tfrac{1}{2}B^2/A,$$

W_p being the polaron energy; we find the hopping energy

$$W_H = \tfrac{1}{2}W_p. \tag{6.4}$$

The hopping energy is thus one-half of the energy released when a polaron is formed. For this so-called adiabatic case, when the electron goes backwards and forwards several times during the period of excitation, the chance per unit time that the electron will have moved from one site to another after the system has relaxed is

$$\omega \exp(-W_H/k_B T)$$

where ω is the attempt-to-escape frequency. The diffusion coefficient D is thus

$$D = \tfrac{1}{6}\omega a^2 \exp(-W_H/k_B T),$$

and the mobility μ, from the Einstein relation $\mu = eD/k_B T$, is

$$\mu = \tfrac{1}{6}(ea^2\omega/k_B T)\exp(-W_H/k_B T). \tag{6.5}$$

For the non-adiabatic case we have according to Holstein to replace ω by

$$\pi^{1/2}I^2 I\hbar(W_H k_B T)^{1/2} \tag{6.6}$$

where $I = I_0 e^{-\alpha R}$ is the transfer integral between the two sites at a distance R apart.

Hopping conduction according to formula (6.5) is predicted only if $T > \tfrac{1}{2}\Theta_D$, where $k_B\Theta_D$ is the quantized energy $\hbar\omega$ of the vibration (phonon) corresponding to the displacement of Fig. 6.1. At low temperatures the polaron behaves like a free particle with greatly increased effective mass m_{eff} given by

$$m_{\text{eff}} = m_0 \exp(W_H \tfrac{1}{2}\hbar\omega). \tag{6.7}$$

We may say that the zero-point energy $\tfrac{1}{2}\hbar\omega$ allows the transfer of the carrier from one of the pairs of Fig. 6.1(b) to the other. The quantity m_0 is given in the adiabatic case by

$$m_0 = \hbar/2\omega R^2.$$

In crystalline materials a characteristic of polaron motion is a difference between E_σ, the activation for conduction, and E_S, that for the thermopower written as $S = (k_B/e)(E_S/kT + \text{const})$. We expect that $E_\sigma = E_c - E_F + W_H$ and $E_S = E_c - E_F$, where E_c is the extremity of the band in which the carriers move. An example is given later in this chapter (Fig. 6.4).

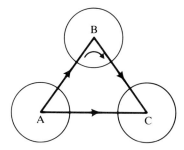

Fig. 6.3. Illustrating the two paths between two sites A, C, for polaron motion, which can give rise to a Hall effect.

Finally we mention the important predictions of Friedman and Holstein (1963) for the Hall effect when charge transport is by polaron hopping. At first sight it is not obvious that, for hopping transport, there should be any Hall effect; but the authors quoted show that, if we take three sites as in Fig. 6.3 so that a carrier can go from A to C either directly or via B, and if thermal activation is such that all three sites have the same energy, there is an interference between the two paths, leading to an activated Hall mobility. An important conclusion is that the sign of the Hall effect is negative (n-type), whether the carriers are electrons or holes so long as they are in s-like orbitals. Emin (1977a,b) discussed what happens if they are p-like; under certain conditions, electrons are then predicted to give positive, and holes negative, values of the Hall coefficient.

The analysis of Friedman and Holstein shows that, in the non-adiabatic case, the Hall mobility μ_H varies with temperature as

$$\mu_H = T^{-3/2} \exp(-\tfrac{1}{3} W_H / k_B T).$$

For the adiabatic case the activation energy can be smaller.

6.3. Polarons in ionic crystals

The analysis given above is for polarons in a crystalline *element*: there is then a sharp condition for polaron formation to occur. In ionic solids this is not so; for a carrier there will always be some deformation of the surroundings, so some self-trapping is always possible, as was first pointed out by Landau (1933). The argument is as follows. Consider an electron placed on one of the cations in a crystal such as KCl. If the ions are all held rigidly in position, the potential energy of another electron at a distance r would be

$$V(r) = e^2 / \kappa_\infty r$$

where κ_∞ is the high-frequency dielectric constant. If the ions are allowed to relax into their new positions in equilibrium, the potential energy becomes

$$V(r) = e^2/\kappa r$$

where κ is the static dielectric constant. So the displacement of the ions produces a potential energy function

$$V(r) = -e^2/\kappa_p r \tag{6.8}$$

where κ is the static dielectric constant. Thus the displacement of the ions produces a potential energy function

$$\frac{1}{\kappa_p} = \frac{1}{\kappa_\infty} - \frac{1}{\kappa}.$$

The electron 'digs its own hole' by polarizing the medium. And because (6.8) describes a Coulomb field, stationary states *always* exist in the potential well. Therefore, if as here we assume the validity of the Born– Oppenheimer approximation, electrons will always be trapped.

A polaron radius r_p can be introduced as follows. We write for the potential produced by the trapped electrons

$$\begin{aligned} V_p(r) &= -e^2/\kappa_p r && r > r_p \\ &= -e^2/\kappa_p r_p && r < r_p, \end{aligned}$$

calculate the energy of the polaron, and choose r_p to minimize it. This calculation was carried out by Fröhlich (1954) and Allcock (1956), who found

$$r_p = 5\hbar^2 \kappa_p/me^2.$$

If, for example, $\kappa_p = 10$, r_p is 25 Å.

If r_p calculated in this way becomes comparable to or less than the lattice constant, this approximation breaks down. A treatment by Bogomolov *et al.* (1968), in which the polarization well is analysed into the normal modes of the lattice vibrations, gave

$$r_p = \tfrac{1}{2}(\pi/6N)^{1/3} \tag{6.9}$$

where N is the number of cations per unit volume.

When r_p becomes large enough, the Born–Oppenheimer approximation breaks down and a dynamic approach must be used, leading to the conception of a 'large' or 'Fröhlich' polaron. On this there is a large literature (see for instance Devreese 1972). In this book we consider only the 'small' polarons for which eqn (6.9) is valid. The same analysis can be used as for the acoustic polaron if we introduce a parameter q $(0 < q < 1)$ to define the potential well $V(r)$ as

$$V(r) = -e^2 q/\kappa_p r.$$

We then find the hopping energy W_H to be equal to $\tfrac{1}{2}W_p$ as before.

There is a clear analogy between the potential well described above and the well formed by the hydration of an ion in solution. For the transfer of electrons between ions in solution to electrodes, metallic or semiconducting, a theory has been developed, notably by Marcus, Dogonadze, and Gerischer, which is very similar to polaron theory (for references see for instance Morrison 1980 and Cannon 1980). The electron is described as 'having a fluctuating energy level in solution'; this means that through thermal fluctuations in the solution shell, the electronic level varies and may reach the Fermi level in the (metal) electrode, in which case the transfer of an electron can occur.

Detailed calculations have been given by Emin of the effect of acoustic and optical phonons separately on hopping motion of polarons. A summary of his work and that of his colleagues is given in Emin (1975) and Gorhan-Bergerin and Emin (1977).

6.4. Observations of small polarons

The most direct way of observing the activated mobility characteristic of polaron behaviour is a measurement of the drift mobility of an injected particle, which should show above a temperature of about $\frac{1}{2}\Theta_D$ an activation energy W_H. Here it is of course possible that a trap-limited mobility will lead to similar behaviour. Some examples of these observations follow. One of the first measurements was that by Adams and Spear (1964) and Gibbons and Spear (1966) on orthorhombic sulphur, a lattice consisting of puckered S_8 rings. (For a review see Spear 1974.) The electron mobility showed an activation energy of 0.24 eV between values of $10^3/T$ of 2.5 and 5. Dolezalek and Spear (1970) showed that the mobility increased exponentially with pressure, suggesting that hopping occurred between weakly overlapping rings and that polaron motion is of the non-adiabatic kind (eqn (6.6)), because for this the factor I^2 is present, depending on $\exp(-2\alpha R)$ and thus sensitively on pressure. This is absent in the adiabatic regime (eqn (6.5)). Activated mobilities are found either for electrons, holes, or both in solid N_2, O_2, and CO (Loveland et al. 1972). In the liquid rare gases (except helium and neon) the electron mobility is high in both solids and liquids (Le Comber et al. 1975), as also for CH_4 (methane), a result for liquids discussed further in Chapter 8; for holes, on the other hand, in neon and argon the mobility is definitely activated, so that self-trapping must occur. The self-trapping is usually thought to be of the type illustrated in Fig. 6.1, the self-trapped hole being called a V_K centre. In alkali and silver halides the hole leads to a homopolar bond between two halogen ions, so that a V_K centre can be described (in chlorides) as Cl_2^-.

On the other hand the possibility exists in metallic halides that an elastic deformation around a single ion (or a pair) traps the hole on that ion (Song 1969). Electron spin resonance signals in AgCl have in fact revealed Ag_2^{2+} ions in mixed crystals of AgCl and NaCl (Olin *et al.* 1984).

The unactivated mobility predicted for polarons when $T \ll \Theta_D$ has not been observed. If it were, for injected particles, it is likely that the polaron bandwidth B_p would be smaller than kT. If so, the mobility should be of the form

$$\mu = elB_p/\{(m_p k_B T)^{1/2} k_B T\} \qquad (6.10)$$

where l is the mean free path. The factor $B_p/k_B T$ arises because in the upper part of the band, which is highly populated, the particle has negative mass. For a high effective mass, however, small fluctuations of potential will trap the particle as explained in Chapter 3, so it is unlikely that an unactivated mobility will be observed. What has, however, been observed is the disappearance of the polaron hopping energy in doped NiO between two localized states (traps); see Busman and Crevecoeur (1966) and Fig. 6.5.

Another way of observing polarons is when, in a semiconductor, the activation energies E_σ for conduction and E_S for the thermopower differ; we can then assume that

$$E_\sigma - E_S = W_H$$

though in non-crystalline and particularly in inhomogeneous materials there are other ways of explaining a difference (cf. Chapter 7). A particularly elegant example is $LiNbO_3$ (slightly reduced), for which conductivity, thermopower, and Hall mobility were measured by Nagels *et al.* (1974). E_σ and E_S differ, and the activation energy for the Hall mobility is found to be one-third of the difference, as expected for polarons. These results for σ and S are reproduced in Fig. 6.4.

There has been some controversy about whether E_σ and E_S are identical for Li-doped NiO, in which the carriers are probably 3d holes in the nickel d-band. Thus Bosman and Crevecoeur (1966) found the activation energies equal at high temperatures, as shown in Fig. 6.5; the behaviour at low temperatures is due to the current being carried in an impurity band. Keem *et al.* (1978) found a difference in high-purity NiO, in which the acceptors are nickel vacancies. They supposed that the presence of lithium in the doped specimens produced strains, which could broaden the 3d band enough to prohibit polaron formation.

Since 1980 there has been considerable discussion of the band structure of NiO, summarized in the book by Tsuda *et al.* (1990). Optical and electrical measurements show a band gap of $\sim 4\,\text{eV}$ but calculations (Sawatzky and Allen 1984) have found it hard to reproduce this quantity; it changes little at the Néel temperature, however, so it does not depend on the ordering of the moments. In any material such as NiO one has to ask whether the

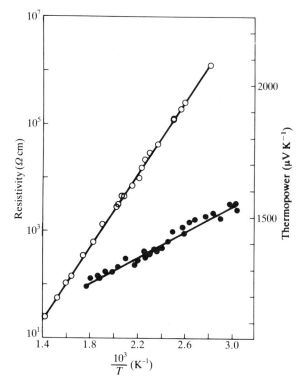

Fig. 6.4. Temperature dependence of the electrical resistivity ($\bigcirc$) and thermopower in μV K^{-1} ($\bullet$) of a single crystal of slightly reduced LiNbO$_3$ (Nagels *et al.* 1974).

coupling between moments is of the Mott–Hubbard type, depending on t^2/U, or the charge transfer type; NiO is probably of intermediate type. The study of lithium-doped NiO by Van Elp (1990) using photoemission spectroscopy and other techniques shows that the holes are in an oxygen band rather than Ni 3d^7; the latter would be expected for a Mott–Hubbard system. Whether the holes form spin polarons is not known. It appears then that the optical edge is caused by a transition leaving a hole in the oxygen 2p band and an electron in a state of Ni 3d^9 type. Whether the activation energy for conduction differs from that for the absorption edge, as one would expect if an exciton is formed, is uncertain. It is thought that at high temperatures the carriers move in a valence band, but at low temperatures in an impurity band, with greatly increased effective mass but smaller activation energy. This is the cause of the drop in S and in the activation energy for conduction. Polarons must be formed here.

In MnO Crevecoeur and de Witt (1968) found a difference in the two activation energies E_S and E_σ, suggesting polaron formation.

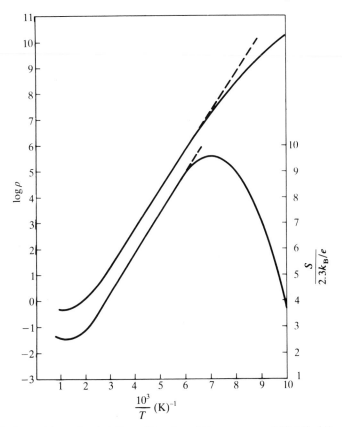

Fig. 6.5. Logarithm of resistivity ρ (Ω cm) and thermopower ($S/2.3(k_B/e)$) of nickel oxide doped with 0.088 per cent of Li_2O (Bosman and Crevecoeur 1966).

UO_2 is a material which shows extrinsic p-type conduction below $\sim 1000\,°C$, intrinsic above. The carriers are U^{3+} and U^{5+}, and both appear to form small, or intermediate, polarons (Harding *et al.* 1980). The material has attracted intensive study because of its importance in reactor design.

Apart from the case of $LiNbO_3$, at the time of writing we know of no measurements for crystalline materials of the Hall effect for polarons, and no evidence for an anomalous sign for holes.

Since polarons are more likely to form in materials for which the band for a given carrier is narrow, organic molecular crystals would appear to be strong candidates; the transfer integral usually lies between 10^{11} and $10^{13}\,s^{-1}$. Since an early paper by Lyons (1957) the conduction mechanism has been extensively discussed. Recent descriptions are by Pope and Swenberg (1982) and the reviews by Silinsh (1980) and by Bassler (1982); it appears that the conditions for (molecular) polaron formation do not usually occur (Vilfan 1973). However, interaction with phonons is strong enough for scattering to occur each time an electron jumps from one molecule to another. Silinsh and Jurgis (1985) used a model of a 'nearly small polaron' in their interpretation of the conduction mechanism.

Another class of material in which a narrow band is expected is the ferrite Fe_3O_4 and compounds such as $PbFe_{12}O_{14}$ investigated by Pollert *et al.* (1983). Here carriers are most easily formed by producing from two Fe^{3+} ions an Fe^{4+} scarcely mobile, and an Fe^{2+}, for which the motion from ion to ion gives a comparatively wide band with a density of states proportional to $\sqrt{E}$, as shown by the behaviour of the photocurrent i, for which $(ihv)^{1/2}$ is proportional to $hv - E_G$. Here v is the frequency of the radiation and E_G the energy of the gap. In non-stoichiometric samples Fe^{2+} ions give donors near midgap. The thermopower is almost identical of T down to 100 K while the conductivity has an activation energy of ~ 0.05 eV; this suggests polaron formation and a high density of donors.

6.5. The rate of formation of polarons

The barrier between the free state of a particle and the self-trapped state, for acoustic polarons, as deduced from Fig. 6.2, was estimated for alkali halides by Mott (1978*d*) and for solid rare gases in some detail by Fugol (1978); the barriers' heights are small (~ 0.02 eV). Sumi (1972) suggested that materials may exist in which there is an equilibrium between free and self-trapped carriers; thus if n electrons per unit volume are in the conduction band and of these x are untrapped

$$x/(n - x) = (N_{\text{eff}}/N) \exp(- W_{\text{p}}/k_B T).$$

Here N is the number of sites per volume, and $N_{\text{eff}} = (2\pi m k_B T/h^2)^{3/2}$. It was suggested that such a model might be applicable to measurements of the electron mobility in anthracene and naphthalene, for which the observed behaviour, a mobility along the c-axis roughly independent of temperature, was described by Schein and McGhie (1979) and Duke and Meyer (1981). However, Sumi (1978, 1979*a,b*) later considered that this hypothesis must be rejected, and proposed, for movement perpendicular to the plane of the molecules, that there is a hopping process essentially activated by fluctuations in the distance between molecules. Sumi's argument is essentially as follows. At very low temperatures, in a crystalline material a band model must be appropriate. In these materials, the transfer integral I_c along the c-axis, that is perpendicular to the planes of the molecules, is much smaller than in the planes. Thus the mobility will be low along the c-axis, but will increase with decreasing temperature. However, for these materials I_c is thought to be low, partly because of the cancellation of two integrals, and small rotations of the c-axis will greatly increase it. Thus at high temperatures, at each molecular vibration, a configuration occurs, for a time proportional to T, which gives a large probability of the transfer from one layer to the next along the c-axis. This should thus give a diffusion coefficient proportional to T for $k_B T \gg \hbar\omega$, and by the Einstein relation a mobility independent of T.

Measurements of the height of the barrier impeding polaron formation for holes or electrons, which should lead to a delay in the self-trapping process, are of interest; a delay has been observed in reaching the final low value of the mobility of holes in vitreous SiO_2 (Hughes 1977) which Mott (1978d) suggested may be due to the process of Fig. 6.1 with a barrier height of 0.2 eV; other interpretations are possible. More recently, Laredo *et al.* (1981, 1983) made a direct measurement of the rate of self-trapping of holes in AgCl, and found an activation energy of 1.8 meV, less than that calculated by Mott. Migration of self-trapped holes was found to be athermal below 30 K; above 35 K the hole moves with a diffusivity $7 \times 10^{-7} \exp(-61 \text{ meV}/k_B T) \text{ cm}^2 \text{ s}^{-1}$. It is deduced that the bandwidth for the self-trapped hole is ~ 2 meV, while the energy to free it into the conduction band is 0.1 eV.

A similar form of self-trapping, with a delay, is shown by excitons in alkali halides and solid rare gases. The exciton is envisaged as a hole, trapping an electron in a fairly extended orbital. Thus Kuusmann *et al.* (1976) observed for NaI both the luminescence of the self-trapped exciton at 4.2 eV and the band-edge luminescence at 5.55 eV characteristic of the unrelaxed exciton. The temperature dependence of the luminescence indicates a barrier of ~ 0.02 eV. Excitons are thought to migrate several hundreds of lattice constants before self-trapping. The evidence was reviewed by Mott and Stoneham (1977) and Rashba (1982).

6.6. Bipolarons

The concept of the bipolaron was introduced by Schlemker and co-workers (Lakkis *et al.* 1976; Schlenker and Marezio 1980) to account for the properties of the compound Ti_4O_7 over a certain range of temperature; this material undergoes a metal–insulator transition of the Verwey type first investigated for Fe_3O_4; its composition can be written $Fe^{3+}(Fe^{2+}Fe^{3+})O_4$ and below 119 K the ions within the bracket are ordered, while at this temperature a first-order transition to a disordered state occurs, with a marked rise in the conductivity. In Ti_4O_7 the chemical formula demands that, if the oxygens are considered as in the state O^{2-}, there must be equal numbers of Ti^{3+} and of Ti^{4+} ions, that is to say titanium ions in the states $3d^0$ and $3d^1$. Berween about 130 and 150 K a phase exists in which the

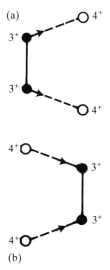

Fig. 6.6. Mechanism for the movement of a bipolaron in Ti_4O_7. (a) Two electrons of a Ti^{3+} pair simultaneously leave their site towards an adjacent Ti chain. (b) Another Ti^{3+} pair is formed on an adjacent chain (Lakkis *et al.* 1976).

resistivity is slightly activated (it varies between 1 and 10^2 Ω cm) but in which the susceptibility is diamagnetic; the charge carriers are believed to be pairs of electrons on adjacent lattice sites, the two forming a homopolar bond with each other and displacing the Ti atoms on which they sit towards each other. The mechanism by which they move is illustrated in Fig. 6.6.

The concentration of these bipolarons is of course high. For the existence of bipolarons formed from excited electrons in semiconductors, the formation energy would have to be nearly as great as the excitation energy for there to be an appreciable concentration. An example where this appears to occur is WO_{3-x} with x of the order 0.001, so the concentration of carriers is small. Schirmer and Scheffler (1982) and Gehlig and Salje (1983) showed that the carriers are paired diamagnetically, with an activation energy for transport of 0.18 eV and a binding energy greater than 0.36 eV. They can be dissociated by radiation in a broad band of energies peaked at 1.1 eV. The binding energy may be less than the energy of pairing. Onada *et al.* (1982) also found evidence for their existence in the quasi-one-dimensional conductor β-$Na_{0.33}V_2O_5$.

A general discussion of bipolarons and their relation to Cooper pairs and electron–phonon interaction leading to superconduction was given by Alexandrov and Ranninger (1981) and will be discussed further in Chapter 12.

Quite recently the concept of a bipolaron has proved applicable to materials formed from long chains (particularly polyacetylene) which when doped show quite high conductivity. A single charge is shown to form a polaron, and a pair of charges a non-magnetic bipolaron (see Klipstein *et al.* 1985; Brazovskii and Kirora 1984).

A model in which pairs of electrons hop simultaneously was proposed by Elliott (1977, 1980; see also Davis 1984) to describe the a.c. conductivity of chalcogenide glasses, materials which are described in the next section. Chapter 10 describes their possible role in high-temperature superconductors.

Salje and co-workers have considered materials in which a high concentration of carriers are available, but only a proportion of them are capable of forming bipolarons because of the 'overcrowding effect'; there is not room for them. Salje and Guttler (1984) have examined the case of WO_{3-x}. Spin pairing leads to the formation of bipolarons with a pairing energy of 1.0 eV (Gehlig and Salje 1983), for low concentrations all carriers form bipolarons, but above 3.5×10^{21} cm some remain as electrons. We think these materials are not superconductors because the carriers are too heavy. We discuss this concept for high-T_c superconductors in Chapter 10.

6.7. Polarons in non-crystalline materials

There is an extensive literature on polarons in amorphous materials. Near the mobility edge E_c in a non-crystalline semiconductor, an electron with energy below E_c, since it is localized, will necessarily distort its surroundings. Whether it will do so to any appreciable extent for energies above the mobility edge is open to question; a discussion is given in Chapter 7 where, for amorphous silicon, it is supposed that it does not, though Emin (1977a,b) in his discussion of the anomalous sign of the Hall effect made an opposite assumption. A comprehensive treatment for energies above and below E_C was given by Thomas and co-workers (Thomas 1985; Müller and Thomas 1984; Fenz *et al.* 1985).

In our view, for certain narrow-band materials in which at zero temperature the band is partially occupied, it must be a good approximation at low temperatures to consider the carriers as small polarons in which the mass is enhanced according to eqn (6.7), so that localization occurs for much smaller values of the disorder than would otherwise be the case. The carriers in the conduction band are then to be considered as a degenerate gas of small polarons. The compound $La_{1-x}Sr_xVO_3$ may be a case in point. A plot of conductivity against $1/T$ (Dougier and Casalot 1970; Dougier 1975) is shown in Fig. 6.7. It is suggested that disorder is produced by the random positions of the ions La^{3+} and Sr^{2+}. The results show that this is sufficient to localize the electron states at the Fermi level when $x < 0.2$. Conduction will then be by excitation to a mobility edge in the d-band (at lower

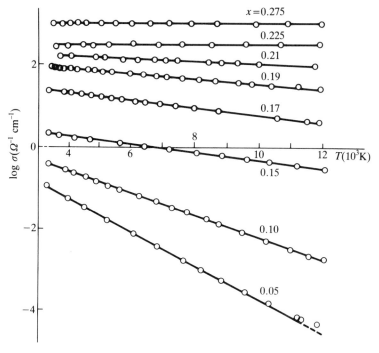

Fig. 6.7. Log (conductivity) of $La_{1-x}Sr_xVO_3$ as a function of $1/T$ for different values of x (Dougier 1975).

temperatures Sayer *et al.* (1975) find $T^{1/4}$ hopping). The small activation energies (~ 0.1 eV) suggest a very narrow d-band, with localization by the random field of the ions. This could only occur for large effective mass, which we ascribe to polaron formation.

Polarons should be strongly and inelastically scatted by phonons ($L_i \sim a$) as the temperatures reaches $\sim \frac{1}{3}\Theta_D$, where Θ_D is the Debye temperature. It seems then reasonable to expect that the pre-exponential factor in the conductivity should be $0.03e^2/\hbar a$ (cf. eqn (3.11)). The value deduced from Fig. 6.7 is about $1000\ \Omega^{-1}\ cm^{-1}$ which is somewhat larger.

We have suggested here that polaron formation narrows the band, which should certainly be so in the metallic regime. For the localized regime, the opposite may be the case, because the Stokes shift will be small for weak localization, large when it is strong.

Emin (1974) in a detailed discussion of polaron behaviour in non-crystalline solids points out that below the Debye temperature the conductivity can often be represented, over a certain range of temperature, by an equation of the form $\exp(-A/T^n)$, similar to that deduced from the variable-range hopping process.

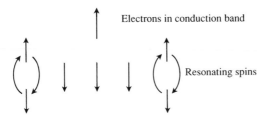

Fig. 6.8. Showing motion of a spin polaron in an antiferromagnetic lattice.

6.8. Spin polarons

The concept was introduced by de Gennes (1962). If a carrier (electron or hole) is introduced into an antiferromagnetic insulator, it is possible that it orients a number of the localized moments round it as illustrated in Fig. 6.8. If I is the energy gained when the carrier has spin antiparallel to the fixed moment, then in the volume $(4\pi/3)R^3$ of the polaron, the energy gained by a carrier is $\frac{1}{2}I$ throughout the volume $(4\pi/3)R^3$. If this is great enough to trap the electron, the energy drops by cI ($c < 1$), and because it is in a hole of radius R we add a positive energy $\hbar^2\pi^2/2mR^2$. An additional positive energy $(4\pi/3)(R/a)^2J_2$ must be added for destroying the antiferromagnetic order within the polaron. Minimizing the total energy within the polarons, we find

$$R^5 = \pi\hbar^3 a^3/4mJ_2$$

and the total energy is

$$\frac{5\hbar^2\pi^2}{6m}\left(\frac{4mJ_2}{\pi\hbar^2 a^3}\right)^{2/3} - J.$$

As long as this is negative, a spin polaron should be formed.

Tentative theories of spin polarons give it an effective mass of order $m \exp(\gamma R/a)$ with $\gamma \sim 1$.

Vigren (1973) showed that the diffusion coefficient of a spin polaron should be of order $\frac{1}{6}\omega_N a'^2$, where $\hbar\omega_N = kT_N$, T_N being the Néel temperature and a' the distance moved by a spin flip at the periphery of the polaron, where it is argued that any activation energy disappears. a' is the distance that the polaron diffuses as a consequence of a spin flip. We consider this should be valid down to low temperatures.

The spin polaron, in contrast to the dielectric polaron, becomes heavier the greater its radius. It does *not* move by activated hopping motion.

Experimental evidence for the existence of spin polarons comes from the work of Von Molnar *et al.* in $Gd_{3-x}V_xS_4$; V here denotes a vacancy.

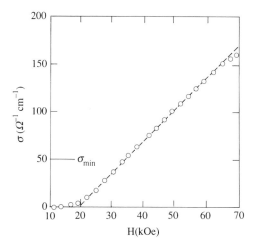

Fig. 6.9. The conductivity of $Gd_{3-x}v_xS_4$ at $T = 300$ mK as a function of magnetic field (von Molnar *et al.* 1985). $\sigma_{min} = 0.03 \, e^2/\hbar a$ is marked.

Electrons in the conduction band are thought to form spin polarons by orienting about 20 Gd moments with anti-parallel spin, as in Fig. 6.9. They are thus heavy, and Anderson localized by the random field of the vacancies. A strong magnetic field, by lining up all the moments in the same direction, destroys the polarons; conductivity then sets in, as shown in Fig. 6.9.

The problem of carriers in antiferromagnetic insulators has been discussed by many authors, such as Methfessel and Mattis (1968) and Nagaev (1983). The second author's 'ferrons' seem identical with our 'spin polarons'. Fields which destroy the antiferromagnetic order at low T have striking effects, such as a large drop in the optical band gap.

Yakovlev *et al.* (1992) have given evidence for the formation of magnetic polarons by excitons excited in a very thin quantum well of CdTe localized by semimagnetic barriers of $Cd_xMn_{1-x}Te$. The idea is that the electron in the exciton penetrates the barrier and aligns some Mn moments antiparallel to itself. This lowers the energy of the exciton when in thermal equilibrium, by 25 meV in a well 6 Å thick. Under magnetic fields of 6 T (at $T = 1.6$ K) the effect disappears.

7 Non-crystalline semiconductors

7.1. Introduction

Whereas in impurity bands in crystalline semiconductors the environment in which the electrons move is known *a priori*, in amorphous semiconductors this is not so. Our understanding of their electrical behaviour depends both on the rules for the motion of electrons deduced from the theories and experiments described in previous chapters, and on investigations of their structure.

We shall discuss in this chapter some of the more important non-crystalline semiconductors and in Chapter 9 materials with large values of the band gap, such as vitreous silicon dioxide and oxide glasses. We must distinguish between materials that can be formed by rapid cooling from the melt, such as vitreous SiO_2 and certain of the chalcogenide glasses, and those that cannot, such as amorphous silicon and germanium.† In the latter, the coordination number is the same as that in the crystal, namely four, but in the melt it is larger, and the liquids are metallic, with properties similar to those of liquid lead. Amorphous silicon and germanium can therefore be prepared only by deposition from some gaseous phase, for instance in a glow discharge, or by bombarding the crystals with fast ions.

Various methods exist for determining the coordination number in glasses and other non-crystalline materials, including X-ray, electron and neutron diffraction, and EXAFS (extended X-ray absorption fine structure). The radial distribution function of amorphous germanium determined by electron diffraction is shown in Fig. 7.1. In general it is found in elements that the coordination number is the same as in the crystal. In glasses with two or more components, such as As_xSe_{1-x} with varying composition, it is normally the case that each As atom (for example) has three neighbours and each Se atom has two, Ge or Si if present 4. This has been called the $8 - N$ rule, meaning that the number of neighbours of each atom is equal to the number of bonds that it can form. It is reasonable to suppose that this is the state of lowest free energy, likely to be formed in a glass obtained by cooling from the melt.

The $8 - N$ rule was postulated by Mott (1969) to account for a property

---

† Very rapid cooling after melting by laser irradiation can produce an amorphous form.

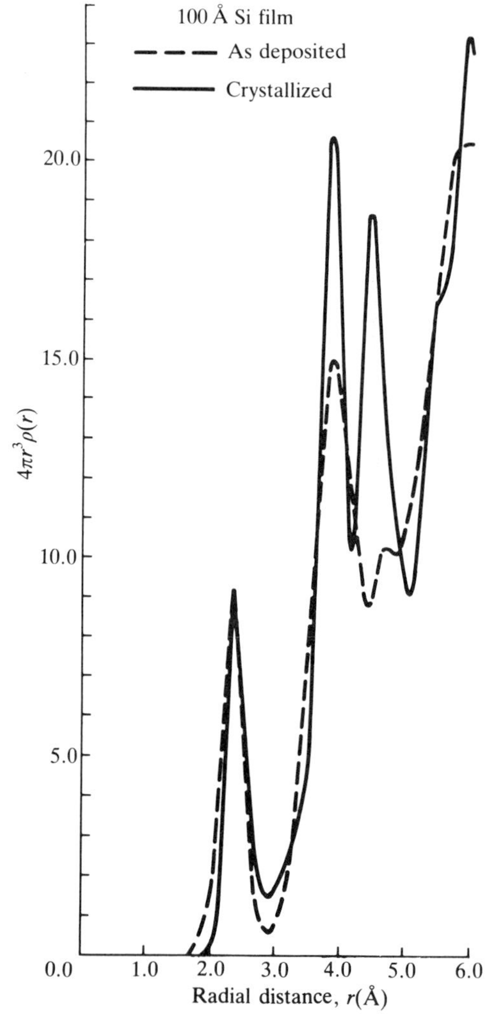

Fig. 7.1. Radial distribution function of amorphous (evaporated) and crystalline silicon determined by electron diffraction (Moss and Graczyk 1970).

of some amorphous semiconductors which is strikingly different from that of crystals. The late B. T. Kolomiets of the Ioffe Institute in Leningrad with his colleagues investigated since the early 1960s the electrical properties of the chalcogenide glasses (see Kolomiets 1964). These include As_2Te_3, As_2Se_3, and glasses which diverge strongly from the stoichiometric composition, formed from Si, Te, As, and Ge (STAG glasses). These materials have a band gap of about 1 eV upwards and so are transparent in the red or near-infrared.

The conductivity varies approximately as

$$\sigma_0 \exp(-\tfrac{1}{2}\Delta E/k_{\mathrm{B}}T)$$

where ΔE is near to the magnitude of the gap obtained optically. In this respect they thus behave like intrinsic semiconductors. The new property established by Kolomiers was that they cannot be doped. Small changes of composition produce only a small change in the conductivity, while crystallization can increase it by orders of magnitude. If all electrons from the impurities are accommodated in bonds, then clearly there will be no loosely bound electrons which can be released into the conduction band by thermal energy. This is in sharp contrast with the behaviour of crystalline silicone doped for instance with phosphorus, in which four of the outer electrons of the latter element form bonds, and one is weakly bound.

We believe the $8 - N$ rule to be generally satisfied for glasses, but not necessarily for deposited films. Thus, as first shown clearly† by Spear and Le Comber (1975), amorphous films of silicon can be formed doped with phosphorus; some of the latter element is incorporated in the network with three neighbours, so that, of the five outer electrons, two are in s-orbitals and inactive and three form bonds, so the atom does not act as a donor. Some, however, go in with fourfold coordination as in the crystal, and these are electrically active.

We see then that in amorphous semiconductors conduction and valence bands and an energy gap are present, just as in crystals. Important work by Grigorovici and Manaila (1970) and Grigorovici (1980, 1983) on bonding in those materials is summarized by Grigorovici and Gartner (1985). The properties of the materials, however, differ from those of crystals also in the following ways.

1. Structural defects, where the normal coordination is not fulfilled, produce states in the gap, which have a major effect on the electrical properties, and which determine the position of the Fermi energy.
2. In the conduction and valence bands the disorder produces tails of localized states, which may affect the drift mobility.

7.2. States in the gap; hydrogenation and doping

As we stated in § 1.1, the Bloch–Wilson theory does not clearly predict an energy gap in non-crystalline materials. In this chapter we shall suppose,

† Earlier work by Chittick et al. (1969) had shown that the conductivity of glow-discharge-deposited amorphous silicon was greatly increased by PH_3 in the discharge.

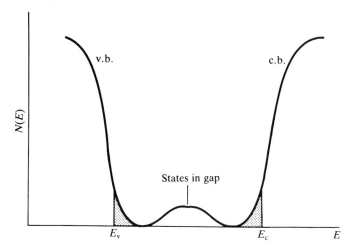

Fig. 7.2. Showing the assumed density of states $N(E)$ in an amorphous semiconductor. E_c, E_v denote the mobility edges; v.b. is the valence band and c.b. the conduction band.

following the considerations of § 3.2 and Anderson's work of 1958, that conduction and valence bands and a gap exist, that in the conduction bands the disorder produces a tail of localized states, and that somewhere within this tail there is a mobility edge, as in Fig. 7.2. In hydrogenated amorphous silicon the range ΔE of the energies where states are localized is thought to be about 0.1 eV, though if an *exponential* tail exists, as now seems probable and was first predicted by Halperin and Lax (1966), the bottom of the band is rather ill-defined.†

There are also deep states in the gap. The current model for amorphous silicon and germanium is that most of these are caused by dangling bonds. A dangling bond is illustrated in Fig. 7.3. It occurs when a silicon atom is so placed that it can form only three bonds with neighbours. If the fourth s–p orbital is singly occupied, as will be the case for a neutral centre, it will give an e.s.r. signal, and such a signal is observed with $g = 2.0055$; g is here the so-called g-factor determining the observed moment. The dangling bond can act as a deep donor, and also as an acceptor, because an electron can

† Perhaps the strongest evidence for the existence of the exponential tail comes from the work of Tiedje and Rose (1980) and the interpretation by Orenstein and Kastner (1981) on the rate of decay of photoluminescence in a-Si–H. With an exponential density of states of the form $N(E) \sim \exp(E/k_B T_0)$, the decay should behave like $t^{-(1+\alpha)}$ where $\alpha = T/T_0$; agreement with this expression was obtained. Theye *et al.* (1985) found in amorphous sputtered or evaporated germanium that the tail, as judged by its effects on the optical absorption, decreases on annealing but that the gap states do not. Thus the tail must be due to strains rather than to defects. For a further discussion of the exponential tail see Davis *et al.* (1985).

Fig. 7.3. A dangling bond in a random network simulating the structure of amorphous germanium or silicon.

be excited from the valence band to form a double occupied state, denoted by D^-. Because of the random environment, the energy level of an electron in the D^- state varies from one site to another. Although the states will be strongly localized in the sense of Anderson (1958), one can think of a band of energies; this is labelled D^- in Fig. 7.4, which shows the resulting density of states. Also, if one of the dangling bonds is unoccupied, the energy will again vary from site to site and can be represented by a band, labelled D^0 in Fig. 7.4. The difference in the energies of the peaks of these two bands is the Hubbard U,

$$U = \langle e^2/\kappa r_{12} \rangle,$$

giving the average repulsion between two electrons in the site. It appears to be surprisingly small, 0.4 eV for Si and 0.1 eV for Ge (Stutzmann and Stuke

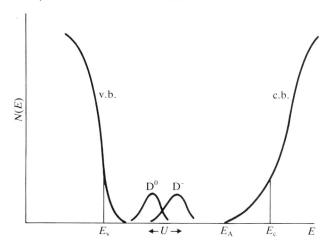

Fig. 7.4. Schematic density of states in amorphous silicon or germanium. E_c, E_v are the mobility edges in the conduction band (c.b.) and valence band (v.b.), respectively, E_A is the bottom of the conduction band, and D^0, D^- are dangling bond states for the first and second electron to occupy the state. This is *not* a 'one-electron' density of states, D^0 existing for an added electron when the gap states are unoccupied, D^- when they contain one electron each.

1983). An experimental determination by Vardeny and Taux (1985) gave 0.5 eV for a-Si–H.

All the evidence suggests that these two bands overlap. Therefore, at zero temperature, some D^- states are occupied. This has the result that the density of states at the Fermi energy $N(E_F)$ is finite. Consequences of this are:

1. Variable-range hopping should be the dominant transport mechanism at low temperatures; this has been observed.

2. Weak doping will shift the Fermi energy only slightly.

3. Electrons and holes can recombine by making transitions to states at the Fermi energy.

Since (3) is likely to be the predominant mechanism for recombination, for solar cells and photoconductive devices it is important that the concentration of dangling bonds should be made as small as possible. In their pioneering work on the production of films of amorphous silicon from a glow discharge in silane (SiH_4), Spear and Le Comber (1972) found excellent photoconductive properties, resulting from the long lifetimes. This was later shown† to be caused by incorporation of hydrogen in the film, as a

---

† Paul *et al.* (1976) introduced hydrogen into sputtered films, thereby decreasing the conductivity; Triska *et al.* (1975) showed that hydrogen could be driven off. Paul (1985) gave a history of the various contributions made to our understanding of the role of hydrogen.

consequence of which a high concentration of Si–H are formed which are passivated by the hydrogen, and the number of unpassivated dangling bonds is greatly reduced. More recent work showed that dangling bonds can be passivated by hydrogen introduced in other ways. Nearly all recent work is thus on hydrogenated amorphous silicon, much of it produced by deposition from a silane glow discharge.

In 1975 Spear and Le Comber showed that amorphous silicon deposited from a glow discharge *could* be doped, by including PH_3 (n-type) or $(BH_3)_2$ (p-type) in the discharge. It was thus possible to form a p–n junction. Under all conditions, in for instance the case of n-type doping, some of the phosphorus will go in with threefold coordination so that it is inactive; this is what is to be expected from the $8 - N$ rule. But some goes in with fourfold coordination as in the crystal, producing thereby a donor. The electron, however, does not remain in the donor but its energy drops to the Fermi energy E_F, since $N(E_F)$ is finite. The conductivity remains intrinsic, but the activation energy $E_c - E_F$ decreases with the degree of n-type doping. Up until now it has not proved possible, as it is for crystals, to move the Fermi energy into the non-localized parts of the conduction or valence band donor. Stutzmann and Street (1985) first observed an e.s.r. signal from these states, in heavily compensated material.

At first sight it looks as if the $8 - N$ rule is violated, but Street (1982) suggested a sense in which it is not. The key point is that the efficiency of doping decreases with the concentration of PH_3 in the discharge, and that the concentration of dangling bonds increases. Street supposes that fourfold coordinated phosphorus atoms can occur only if the electron forms part of a negatively charged dangling bond (D^-), which is equivalent to saying that its energy has fallen to the Fermi level. Assuming a thermodynamic equilibrium between the two phosphorus configurations in the reaction

$$P_3^0 \rightleftarrows P_4^+ + D^-$$

then, if N_0 is the total concentration of phosphorus and n that of the fourfold coordinated material, the law of mass action gives

$$(N_0 - n)/n^2 = \text{const}$$

and when the doping efficiency η ($= n/N_0$) is low this leads to the relation

$$\eta = \text{const}/N_0^{1/2}.$$

Figure 7.5 shows that this relationship is satisfied.

Doping of n-type, by increasing the number of doubly occupied states, decreases the e.s.r. signal; p-type doping will first increase it, and then as empty states predominate will decrease it. It is by measurements of the e.s.r. signal that the energy illustrated in Fig. 7.4 between the peaks of the D^0 and D^- bands has been determined.

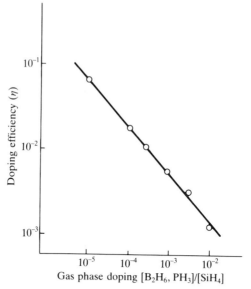

Fig. 7.5. The doping efficiency η of amorphous Si:H for different dopant levels (Street 1982).

7.3. Experimental investigation of the density of states in the gap

Various methods exist for determining the density of states in the gap and in the band tail, which have been applied particularly to hydrogenated amorphous silicon. Of these the 'field effect' has been extensively used. This measures the increased conductivity in the region of the surface when a field F is applied perpendicular to the surface. The field will penetrate to a distance

$$\lambda = \{\kappa/4\pi e^2 N(E_F)\}^{1/2},$$

and so, since the charge per unit area induced by the field is $F/4\pi$, the charge density in the layer is per unit volume

$$F/4\pi\lambda.$$

The Fermi energy is changed by

$$\delta E = e^2 F/4\pi\lambda N(E_F)$$

varying as $\{N(E)\}^{-1/2}$; since the current when ΔE is large is mainly in the surface layer, $N(E)$ can be deduced. This method has established the exponential tails to the density of states of the form $N(E) = A\exp(E/k_B T_0)$ some way below a mobility edge, which is probably related to the random

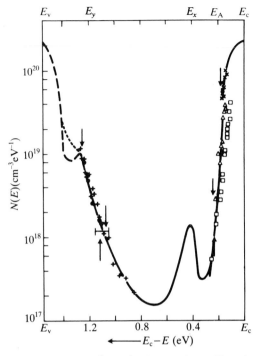

Fig. 7.6. Density of states in $cm^{-3}eV^{-1}$ of amorphous Si:H deposited at a high temperature. The central region refers to undoped specimens and the points to lightly doped. The exponential edge is shown; E_c, E_v are the two mobility edges, the dotted line above E_v is an extrapolation, and E_x and E_y are the positions of the two dangling bond states (adapted from Spear and Le Comber 1984).

lattice rather than to defects. Data from Spear and Le Comber (1984) are shown in Fig. 7.6. Another method entails the use of space-charge limited currents, and a third uses deep-level transient spectroscopy. Unfortunately these methods do not all give the same results, as shown in Fig. 7.7. Fritzsche (1985) gives a discussion of the reasons for this. Of particular importance is the density of states at about 0.8 eV below the mobility edge, as we shall see in the next section.

7.4. Conductivity and thermopower

If a mobility edge exists and charge transport is through electrons with energy just above it, one expects that the conductivity will be of the form

$$\sigma = \sigma_0 \exp(-E_\sigma/k_B T)$$

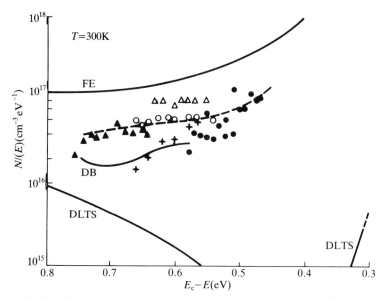

Fig. 7.7. Density of states in glow-discharge a-Si deduced from different experiments. FE, Field effect; DB, space-charge-limited current; DLTS, deep-level transient spectroscopy.

and the thermopower

$$S = (k_B/e)(E_S k_B T + A).$$

If there is a sharp discontinuity in the mobility at E_c, we have seen (§ 3.11) that $A = 1$, and under other assumptions that $A = 2$. The assumption that a mobility edge exists suggests that

$$E_\sigma = E_S = E_F - E_c.$$

However, E_F will vary with temperature for small T according to the equation

$$E_F = E_0 - \frac{\pi^2}{6} (k_B T)^2 \frac{d \ln N}{dE}.$$

Small and rapidly varying values of $N(E)$ near the Fermi energy can give for E_F a behaviour as in Fig. 7.8. If at room temperature one approximates by writing

$$E_F - E_c = E_0 - \gamma_1 T,$$

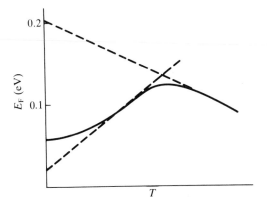

Fig. 7.8. Suggested variation with temperature of Fermi energy E_F in eV.

we see that

$$\sigma = \sigma'_0 \exp(-E_0/k_B T)$$

where

$$\sigma'_0 = \sigma_0 \exp(\gamma_1/k_B)$$

and

$$S = (k_B/e)\{E_0/k_B T + A - \gamma_1/k_B\}.$$

The experimental values of σ'_0 vary strongly with doping; Fig. 7.9(a) shows the values plotted against the Fermi energy due to Overhof and Beyer (1983); Fig. 7.9(b) shows data over a wider range from Heintze and Spear (1986). If the model given here is correct, we can deduce that in a certain range of energies $N(E_F)$ is indeed small and rapidly varying.

The results given in Fig. 7.9(a) are an example of the Meyer–Neldel rule which has a wider application as describing how the observed prefactor varies with activation energy. This is described in detail by Overhof and Thomas (1989). Modern methods of determining the density of states have made it possible to determine γ_1, which make it in principle possible to determine the true exponential $\sigma(E_c)$. This according to the theories of earlier chapters should be given by

$$\sigma(E) = 0.03e^2/\hbar L_i.$$

Mott (1988) in a detailed discussion of the problem writes for the observed prefactor

$$\sigma_0 = \sigma(E_c) \exp\{(\gamma_1 + \gamma_2 + \gamma_3)/k_B\}$$

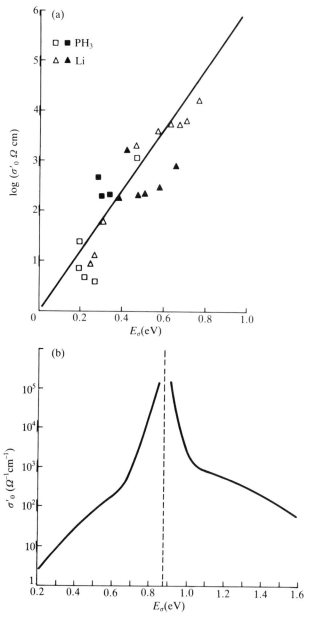

Fig. 7.9. (a) Pre-exponential factor σ_0' as a function of the conductivity activation energy E_σ for phosphorus- and lithium-doped a-Si. Open symbols, low-temperature range; closed symbols, high-temperature range (Beyer and Overhof 1984). (b) As in Fig. 7.9(a), from Heintze and Spear (1986).

where γ_2 defines the change in the band gap with T. γ_3 is defined in the following way.

As pointed out particularly by Overhof and Thomas (1989 and references therein) the mobility edge, though a precisely defined concept at $T = 0$ for a degenerate electron gas, cannot be so defined for single electron in a conduction band, because electrons can always, by emitting phonons or radiation, fall to a state below E_c giving E_c a finite breadth. Instead of a mobility edge, then, the relevant concept is the energy E_M at which $\sigma(E)$ is a maximum. If we write

$$E_M = E_c + \gamma_3 T$$

then according to Thomas (cf. Mott 1988, p. 376), $\gamma/k \simeq 2$.

From all the observed quantities, then, we deduce $L_i \sim 12$ Å for a-Si–H.

A further experimental fact is that, for n-type conduction as determined by the sign of the thermopower, the Hall coefficient is p-type, and has again the opposite sign to that given by the thermopower. Some of these results are shown in Fig. 7.10, where the Hall constant of microcrystalline material is plotted against crystalline size. It will be seen that the anomalous sign appears only for a crystalline size below ~ 20 Å. Emin (1979a,b) gave an explanation of this behaviour, based on his hypothesis that conduction is by polaron hopping rather than by excitation to a mobility edge. Mott (1985c)

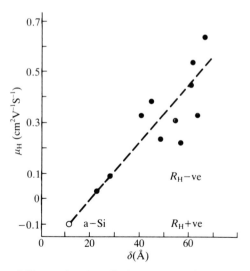

Fig. 7.10. Hall mobility μ_H in glow-discharge amorphous silicon plotted against crystalline size δ. The open circle denotes a typical result for amorphous silicon for which the sign is anomalous (Spear and Le Comber 1982).

has criticized this hypothesis, and gives a tentative explanation in terms of motion at a mobility edge (Mott 1991).

Apart possibly from very perfect specimens, E_S is observed to be smaller than E_σ. The difference depends strongly on the state of the specimen being increased by bombardment. To account for this, Overhof and Beyer (1981, 1983) assumed that there exists a *long-range* random potential in these materials, which is a consequence of charge centres (D^+ and D^-). The thermopower will be determined by the energy in the valleys, the conductivity by the 'cols' from one valley to another.

In early work it was often suggested that the activation energy measured in the drift mobility (see below) was entirely a consequence of such fluctuations of potential (cf. Fritzsche 1960). However, no one has suggested how the anomalous sign of the Hall effect could be described by such a model, and we suggest that both long-range fluctuations *and* a mobility edge exist.

As regards the magnitude of σ_0, this can also be deduced from measurements of the drift mobility described in the next section. From the theoretical point of view, the considerations of Chapter 3 show that we expect

$$\sigma_0 = 0.03 e^2 / \hbar L_i$$

where L_i is the inelastic diffusion length, in this case a consequence of collisions with phonons. For L_i we have

$$L_i = (D\tau_i)^{1/2},$$

where τ_i is the time to such a collision, which will be proportional to

$$\tfrac{1}{2}\hbar\omega + 2k_B T$$

where ω is an appropriate phonon frequency; the factor 2 comes because transitions must be included in which a phonon is emitted or absorbed. Estimates of τ_i lead to values of σ_0 of the order of $16\ \Omega^{-1}\ \text{cm}^{-1}$ (Mott 1985b). As will be seen from Fig. 7.9, the observed value of σ'_0 varies widely about this value. This must be a consequence of the factor $\exp(\gamma_1/k_B)$.

7.5. Drift mobility

Direct measurements of the drift velocity have proved useful in determining the mechanism of charge transport by electrons or holes and give more direct methods of estimating σ_0. Carriers are injected at one point of the sample and the transit time t is measured to another point at a distance d, under the influence of an electric field F. The drift mobility μ_D is then given by

$$\mu_D = d/Ft.$$

The materials on which most experiments have been carried out have typically resistivities greater than $10^7 \, \Omega \, \text{cm}$, so that the dielectric relaxation time $(10^{12} \, \rho\kappa/4\pi)$ is large compared with t; the excess carriers are not screened by other carriers, as in experiments on the drift mobility of minority carriers in crystalline semiconductors (Shockley 1950).

It is supposed that within the pulse electrons are trapped at band-edge states and released several times. Some electrons will be lost to deep states. If the conduction band shows an exponential tail, a clear separation cannot be made between deep and shallow states. An assumption sometimes made is that transitions to the deeper states are improbable, so that effectively a lower limit to the conduction band can be set (see Davis et al. 1985).

Particularly at low temperatures, however, one cannot suppose that an equilibrium is set up within the pulse; the first carriers to arrive at the opposite edge of the specimen are those that have not been trapped, followed by those that have been trapped once, twice, and so on. Transport is then described as dispersive. A theoretical discussion was first given by Scher and Montroll (1975) and there is now a considerable literature on the subject, which will not be reviewed here. One feature of dispersive transport is that the apparent drift mobility depends on film thickness.

If transport is non-dispersive and a lower limit E_A to the conduction band can be assumed, then we may write for the drift mobility μ_D

$$\mu_D = \mu(0) \exp(-\Delta E/k_B T).$$

If μ_{ext} is the mobility for particles with energy E_c, and thus *at* the mobility edge, then

$$\mu(0) = \mu_{ext}\{N(E_c)/N(E_A)\}.$$

The pre-exponential in the conductivity can be written

$$\sigma_0 = eN(E_c)k_B T\mu_{ext}$$
$$= eN(E_A)k_B T\mu(0).$$

Data both from Spear (1983) and Tiedje and Rose (1980) give at room temperature $\mu_D = 0.8 \, \text{cm}^2 \, \text{V}^{-1} \, \text{s}^{-1}$ and $\Delta E = 0.13 \, \text{eV}$, and Spear maintains that transport is non-dispersive. $\mu(0)$ is thus $120 \, \text{cm}^2 \, \text{V}^{-1} \, \text{s}^{-1}$. To obtain σ_0 one needs to make assumptions about $N(E_c)$, $N(E_A)$; using data given by Spear (Fig. 7.6), Mott (1985b)† estimated that σ_0 is about $24 \, \Omega^{-1} \, \text{cm}^{-1}$, a value reasonably close to that calculated in the last section.

It has not been universally assumed that charge transport as usually observed is caused by electrons at a mobility edge. At low temperatures Spear and co-workers in early work considered that conduction at low

† This paper contains a numerical error; see Mott (1986b).

temperatures was by hopping between localized states at a band edge; a theoretical description of the process was given by Grant and Davis (1974). More recently, it was proposed that d.c. conduction is normally in a band tail below a mobility edge. For chalcogenides this was emphasized particularly by Monroe (1985) and by Kastner (1986), who gave a theory according to which a difference between E_σ and E_S can be explained. For hydrogenated amorphous silicon this appears to us improbable; the drift mobility is, according to Spear and co-workers, not dispersive; hopping conduction results in a large magnetoresistance, which is not seen except for the photocurrent at very low temperatures, where conduction must be in the tail; and the difference between E_σ and E_S is highly structure-dependent (though indeed the nature of the tail might be).

The question whether electrons in the conduction band of amorphous silicon are self-trapped as polarons on Si–Si bonds has been debated. It was put forward by Emin et al. (1972) to account for the difference between D_S and E_σ (particularly for chalcogenides); Emin (1977a,b) also gave a theory based on polaron formation to explain the anomalous sign of the Hall coefficient in amorphous silicon and other materials. Cohen et al. (1983) gave arguments to suggest that for energies E above but near E_c, polarons must form before $E - E_c$ vanishes; Mott (1985b) gave arguments suggesting that this is not the case.

7.6. Chalcogenides; valence alternation pairs

The chalcogenide semiconductors such as As_2Te_3 and amorphous selenium have a property different from that of a-Si-H; the Fermi energy seems to be pinned near the centre of the gap, as it is in amorphous silicon, but no e.s.r. signal or other sign of singly occupied states is observed. If the $8 - N$ rule were the whole story, there would be no states in the gap, and the position of the Fermi energy and hence the conductivity would be highly sensitive to any impurity acting as a donor. One can hardly assume that they are totally absent, and yet the Fermi energy is pinned. There seem to be states in the gap, giving a finite value of $N(E_F)$, which do not give rise to uncoupled spins.

The current model to explain this was put forward by Street and Mott (1975) and Mott et al. (1975a) and greatly clarified by Kastner et al. (1976) and subsequent papers. The concept will be illustrated by a discussion of amorphous selenium, in which selenium chains lie parallel to each other. At the end of each chain one would expect a 'dangling bond', containing an unpaired electron. However, the selenium atom contains six electrons outside the closed shell. Two are in s-orbitals and inactive. Two are in p-orbitals and form bonds. Two more are in p-orbitals which do *not* form bonds; they are called 'lone pair orbitals', and the upper part of the valence band can be envisaged as made up from them. So a neutral dangling bond is a site at the end of the chain at which one of these lone pair orbitals contains a single electron. It is now argued that, if an electron is removed from one of the lone pair orbitals of a nearby atom in a neighbouring chain, the two singly occupied orbitals will form a bond; the two atoms will be attracted to each other and this second atom is now threefold coordinated. The resulting defect, positively charged, was called by Street and Mott a D^+ centre and by Kastner et al. C_3^+, denoting the threefold coordination and the positive charge.

If these defects exist in the glasses, or in the liquids from which they are quenched, there must be a corresponding negative centre, the electron must go somewhere. The obvious place for it is in another, perhaps distant, dangling bond, producing the C_1^- or D^- centre. If these centres are to exist in preference to the neutral dangling bonds, it follows that the reaction

$$2C_1^0 \rightarrow C_3^+ + C_1^-$$

must be exothermic. This is sometimes described as resulting from a 'negative Hubbard U', the repulsion $\langle e^2/\kappa r_{12} \rangle$ between the two electrons in the dangling bond being more than compensated by the energy gained as the material contracts locally to form the threefold coordinated selenium. Kastner et al. argued that this must be so, because each broken bond (C_1^-) is always associated with an extra bond formed at C_3^+. The bond energy is high, and structures with uncompensated broken bonds will be avoided.

The neutral centre can be generated by illumination; in liquid Se Cutler (1977) observed it through the e.s.r. signal.

We now ask why the presence of these centres will pin the Fermi energy. We argue that if electrons are excited into a conduction band, *two* electrons will be generated if a C_1^- is transformed into a C_3^+, and, if the conductivity varies as $\exp(-\frac{1}{2}E/k_B T)$, E will be half the energy necessary for this process. But let us suppose that we dope the material. Then the donors will lose their electrons by increasing the ratio of C_1^- to C_3^+ centres. The activation energy E is thus unchanged. The Fermi energy is more firmly pinned than in amorphous silicon.

For the many applications of this concept to photoluminescence, optically induced e.s.r., and other properties, the reader is referred to the books mentioned in the preface and to the review by Street (1976) and his book (Street 1991).

7.7. Photoconduction and recombination

For applications of p–n junctions in amorphous silicon to solar cells it is necessary to maximize the lifetime of electron–hole pairs. These can recombine

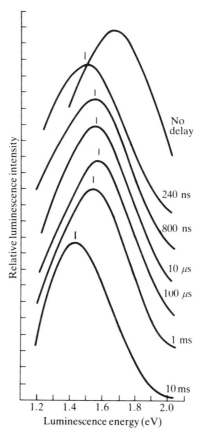

Fig. 7.11. Photoemission bands in a-Si–O–H for various time delays (Street 1980). Excitation energy 2.334 eV; temperature 8 K.

optically, giving rise to photoluminescence, or by emission of phonons. The probability of recombination with emission of phonons decreases rapidly with increasing energy of the transition, which means that the presence of dangling bond states decreases the lifetime. This is why hydrogenated material, or passivation by some other material such as fluorine, is essential to obtain high efficiency.

There has been considerable investigation of the photoluminescence in a-Si–H. The main peak at 1.4 eV was first observed by the Marburg group (Engemann and Fischer 1974, 1976) and is shown in Fig. 7.11, which illustrates measurements of Street (1980) for various delays following excitation. The mechanism proposed is the following (Tsang and Street 1979).

1. The hole is trapped in a shallow level.

2. Rapid recombination is due to electrons which are attracted to the hole, and are not yet thermalized.

3. For times when the rapid recombination ceases, recombination is due to electrons trapped at band-edge localized states. If the electron is near to the hole, the Coulomb attraction will lower the emitted energy. If it is far away at a distance R, tunnelling is important in increasing the lifetime, and a lifetime

$$\tau = \tau_0 \exp(2R/R_0)$$

is expected. Here R_0 is the radius of the wave function of the carrier for which this is largest, and $\tau_0 \sim 10^{-8}$ s. Figure 7.11 shows that emission spectra shift to larger energies for increasing time delays.

Apart from the main peak at 1.4 eV, Street and Biegelson (1980) observed a subsidiary peak at 0.9 eV. This is thought to be caused by electrons trapped at dangling bond states (two electrons in the dangling bond), 0.5 eV below the conduction band. The evidence for this model was described by Mott (1980a, p. 5456).

We now review briefly the mechanism of multiphonon recombination. We describe the energy of an electron–hole pair, whether trapped or otherwise, as a function of some configurational coordinate q. Several possibilities are shown in Fig. 7.12. In Fig. 7.12(a) if q changes after excitation of the system, and comes to equilibrium at the point marked P, radiation can be emitted with energy $h\nu$ given by PQ. The decrease in comparison with the excitation energy AB is called the Stokes shift. Then two things can happen apart from radiation.

1. The system can be thermally excited to X. Recombination can then occur with probability of order

$$\omega \exp(-\varepsilon/kT)$$

where ε is the excitation energy.

2. A direct multiphonon transition can occur. Detailed calculations (Englman and Jortner 1970) show that at zero temperature the probability per unit time for such a transition is

$$\omega \exp(-W\gamma/\hbar\omega).$$

Here ω is the appropriate phonon frequency, W is the energy AP, and γ is in the range 1 to 2. The important point here is that if W is less than about 10 $\hbar\omega$, the probability may be greater than that for optical transitions.

The probability is also increased by temperature; for a discussion see Mott and Davis (1979, Chapter 3).

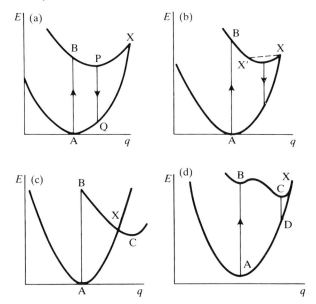

Fig. 7.12. Configurational coordinate diagram for electron–hole pair, showing various possible forms of radiation or (c) trapping.

In Fig. 7.12(c) we illustrate how self-trapping of the excited state at C is possible.

Figure 7.12(d) shows how a barrier may stabilize the state B, leading to a delay as proposed in § 6.5.

In Fig. 7.12(a), if X lies below B, it is possible that after excitation the momentum of the system will bring it to X, so that multiphonon recombination can occur. This was first proposed by Dexter *et al.* (1955). In Chapter 9 we suggest that this must occur in vitreous silicon dioxide.

8 Liquid semiconductors and metal–insulator transitions in liquids

8.1. Introduction

Liquid semiconductors were first studied systematically by the Leningrad school (see the review by Ioffe and Regel (1960) and the books by Gubanov (1963) and Glazov *et al.* (1969)). Recently many of them have been investigated in several other centres. In our view there is little difference in principle between the theories necessary to treat liquid and solid non-crystalline semiconducting materials, except that any defects (such as dangling bonds and valence alternation pairs described in Chapter 7), which may pin the Fermi energy, will have a concentration which minimizes the free energy and which varies strongly with temperature. Cutler and Rasmolon-dramanitra (1984) have analysed the properties of non-metallic Se–Te alloys in these terms; neutral and charged dangling bonds (valence alternating pair) may exist, with motion of charge between them (see also Cutler 1977).

A main theme in this chapter is the discussion of the *liquid* divalent metal Hg, and the monovalent metals Cs and Rb, at temperatures and densities near to that at which a metal–insulator transition occurs, this involving work at high temperatures; we also give a short discussion of metal–ammonia solutions, which also show a transition with decreasing concentration of metal. We normally use a pseudogap model, though in some materials it is preferable to assume that defects exist in thermal equilibrium, for instance, as already stated, neutral and charged dangling bonds, so that hopping or unactivated charge transfer can occur.

The pseudogap model is illustrated in Fig. 8.1. With increasing density of the liquid a full and an empty band of 6s, 6p type in mercury and upper and lower Hubbard bands in the monovalent materials and metal–ammonia, approach and overlap. Were it not for disorder, a band-crossing or Mott transition would occur, and at zero temperature this would show discontinuous change in the number of free electrons from zero to a finite value. But disorder will certainly produce tails which *may* prevent the occurrence of this discontinuity. We have argued in § 4.1 that the discontinuity is smaller in a band-crossing than in a Mott transition, and thus should be smaller in mercury than in caesium or metal–ammonia. Experimentally there is a

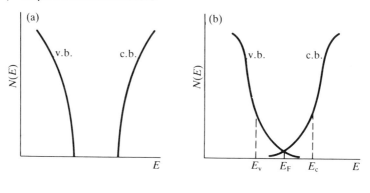

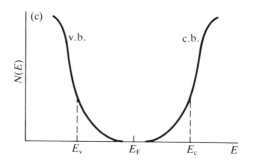

Fig. 8.1. Density of states $N(E)$: (a) for a crystal; (b) for the pseudogap metal discussed here; and (c) for a typical non-crystalline solid. v.b. is the valence band, c.b. the conduction band, E_F the Fermi energy, and E_v, E_c the mobility edges.

striking difference between mercury and caesium. The critical point of the latter is associated with a metal–insulator transition, the conductivity there being of the order of σ_{min}; the same is true of metal–ammonia. On the other hand, at the critical point of mercury the conductivity is of the order of $10^{-3}\ \Omega^{-1}\ cm^{-1}$, some 10^5 times smaller than σ_{min} (Hensel 1990). We have to suppose, therefore, that in mercury disorder wins over the very small discontinuity; the critical point is then caused by forces of the same type as in argon, that is interaction between closed $6s^2$ shells, and the metal–insulator transition, at a much higher density, is of the Anderson type. We should expect a much slower change of structure as the material goes into the insulating state in mercury than in Cs.

In both types, then, in the metallic state we should expect in solids as we approach the insulating regime that the conductivity should be given by the

Kawabata equation (cf. eqn (3.12))

$$\sigma = \sigma_B g^2 \left\{ 1 - \frac{1}{(k_F lg)^2} \left(1 - \frac{l}{L_i} \right) \right\}. \tag{8.1}$$

Mott (1985c) first pointed out that there is considerable experimental evidence that the correcting term is absent in liquids. He argued that in liquids all collisions could be treated as inelastic, so that quantum interference could not occur. Afonin et al. (1987), however, showed that this was not necessarily the case. In this chapter we shall treat in detail the arguments of Mott and Kaveh (1990), which attempt to reconcile the two conclusions. We first consider in more detail the properties of mercury and caesium.

8.2. Fluid mercury

In this divalent metal the 6p bands must overlap; at room temperature any small dip in the density of states (Mott 1972) does not affect the conductivity (see Chapter 3) when the mean free path (7 Å) is considerably greater than the interatomic distance. This is shown by the values quoted in Chapter 2 (Table 2.1). However, by the use of high temperatures and pressures, Hensel and Franck (1968) and others have been able to measure electrical conductivity and thermopower as a function of volume; at constant volume the effects of temperature are comparatively small. The measurements of Hensel and Franck are shown in Fig. 8.2; it will be seen that, as the specific volume increases, the conductivity drops through a value corresponding to σ_{min} well into the semiconducting regime. We have then the possibility of an experimental test on the pre-exponential factor of the consequences of our assumption about the absence of quantum interference, which are that at the transition

$$\sigma = \sigma_B g^2 = 0.03 e^2 / \hbar a. \tag{8.2}$$

The value of the pre-exponential factor factor σ_0 can be obtained experimentally from the thermopower S if we suppose that in the semiconducting regime

$$S = (k_B/e)(E/k_B T + A)$$

and

$$\sigma = \sigma_0 \exp(-E/k_B T)$$

where $E = E_c - E_F$. Then

$$\log_{10} \sigma = \log_{10} \sigma_0 - \{(e/k_B)|S| - A\}/2.3.$$

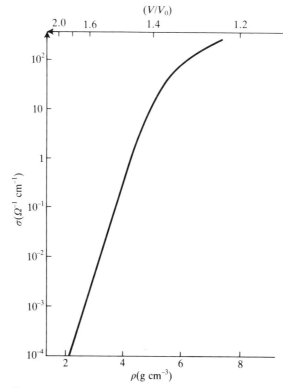

Fig. 8.2. Specific conductivity σ of mercury at 1550°C as a function of density ρ (Hensel and Franck 1968). V is the volume and V_0 the molar volume.

A plot of log σ against S is shown in Fig. 8.5. The slope is exactly as predicted. If we suppose that $A = 1$ (§ 3.5), we can find σ_0 by extrapolating to $S = k_B/e$ ($S = 89 \, \mu V/K$). Schönherr *et al.* (1979) found for conductivities between 200 and $5 \, \Omega^{-1} \, cm^{-1}$ values of σ_0 in the range 140–$200 \, \Omega^{-1} \, cm^{-1}$, which suggests that the correcting term in eqn (3.2) is absent both in the conductivity and thermopower.

There is, however, some doubt as to whether the transition in mercury is of the normal Anderson type. Turkevich and Cohen (1984) proposed a model in which an excitonic insulator is formed. Moreover, there is doubt as to whether $A = 1$ is the correct value and whether for liquids $E_c - E_F$ should contain the temperature variation for both conductivity and thermopower (Emin 1977b; Overhof and Beyer 1983; Butcher and Friedman 1977; Mott 1985d; Butcher 1985).

Results for both liquid caesium (Freyland *et al.* 1974) and liquid selenium are similar, as regards the slope; σ_{min} appears as $300 \, \Omega^{-1} \, cm^{-1}$ for caesium, for selenium rather smaller.

According to El-Hanany *et al.* (1983) the magnetic properties of liquid caesium are similar to those of a ferromagnet above its ordering temperature.

The liquid alloy MnTe is, however, according to Barnes and Enderby (1985) somewhat different. The conductivity is thought to be in a manganese d-band split by the Hubbard U, and thus essentially independent of temperature. Therefore σ_0 can be deduced from the conductivity alone, and is found to be about $200\,\Omega^{-1}\,\text{cm}^{-1}$, as we anticipate.

The $\text{Se}_{1-x}\text{Te}_x$ alloys, investigated by Perron (1967) and Cutler and Rasolondramanitra (1984), appear to provide an example of a liquid in which defects form in thermal equilibrium, giving an impurity band which contributes to the conductivity. According to Cutler and Rasolondramanitra, the main conduction process is in the valence band; $E_v - E_c$ is strongly temperature-dependent, but from measurements of the thermopower and through eqn (8.1), σ_0 can be determined to be about $10^3\,\Omega^{-1}\,\text{cm}^{-1}$ and independent of x. This large value suggests that in the valence band the range of localized states is very small ($\lesssim k_B T$), so that it is without major effect on the scattering, which might be calculated as for a liquid rare gas (Chapter 2). However, a second transport channel is observed, with a high activation energy. In a liquid we may expect both charged dangling bonds at chain ends, *and* neutral ones, both being in thermal equilbrium. The concentration of the latter is determined from e.s.r. measurements. Both neutral and charged centres being present, hopping between them is possible, and a term in the conductivity results which increases rapidly with temperature, as the concentration of defects increases.

Another material in which there is proportionality between σ and g^2 is liquid tellurium; the conductivity, between 1300 and $2750\,\Omega^{-1}\,\text{cm}^{-1}$, is less than the Ioffe–Regel value for a material with six electrons per atom ($6000\,\Omega^{1}\,\text{cm}^{-1}$), implying a pseudogap which gradually fills in with increasing T; this is caused by a gradual increase in coordination number from 2 to 3 (Cabane and Friedel 1971). Figure 8.3 shows that σ is proportional to K^2, where K is the Knight shift. The Knight shift is known to be proportional to the Pauli susceptibility and thus to the density of states. There thus appears then to be strong experimental evidence that, in (8.1) the last term is absent, as conjectured.

8.3. The absence of quantum interference

The evidence for the absence of quantum interference comes from the fact that in many cases the conductivity σ, for changing composition or temperature, obeys the relationship $\sigma^{1/2}$ proportional to g, where g is obtained from the (Pauli) paramagnetism or Knight shift.

An example is provided by the system Te–Tl investigated extensively by

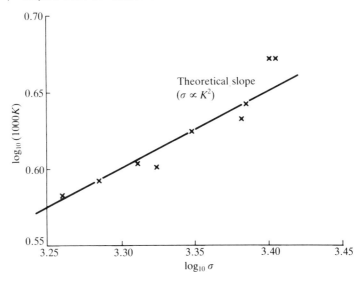

Fig. 8.3. Plot of log $(1000\ K)$, where K is the Knight shift of liquid selenium, versus log σ, showing that $\sigma \propto K^{2.}$

Cutler and co-workers (see Cutler 1977). It is thought that molecules with the composition $TeTl_2$ form in the liquid, and that with liquids in which the concentration of Tl is greater than this there is a degenerate electron gas in a conduction band giving n-type thermopower. For lower concentrations the thermopower is p-type, and the conductivity is lower, though above $100\ \Omega^{-1}\ cm^{-1}$, and thus probabily metallic. The pseudogap model is indicated because a plot of the (Pauli) magnetic susceptibility against the square root of the conductivity, reproduced in Fig. 8.4, shows that

$$\chi = \text{const } \sigma^{1/2} + \chi_0.$$

χ_0 is here the sum of the diamagnetic and Van Vleck contributions. Since χ should be proportional to the density of states, this gives experimental verification of the relationship (8.2), showing that the correcting term in the Kawabata formula (8.1) is absent.

Further evidence is cited by Mott and Kaveh (1990). This includes:

1. The work of Warren (1970a,b, 1972a,b) on the Knight shift K for In_2Te_3 and Ba_2Te_3 in the liquid state. K should be proportional to g and if g is deduced from the conductivity assuming $\sigma \propto g^2$; the proportionality is maintained from $g = 1$ until $g = 0.2$, at which point it is assumed by Warren that localization occurs.

2. The work of Acrivos (1972) and Acrivos and Mott (1972) (see also Mott 1989, 1991, p. 242 et seq.) on metal–ammonia solutions in the range

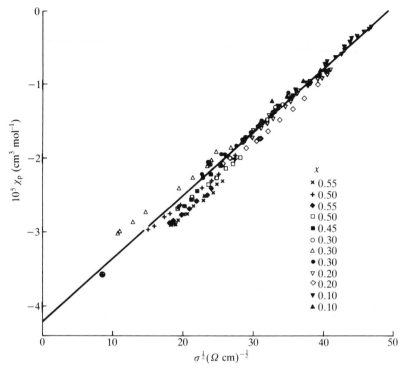

Fig. 8.4. Pauli susceptibility χ_P of $Te_{1-x}Tl_x$ liquid alloys plotted against $\sigma^{1/2}$ for various temperatures and composition (Cutler 1977).

of σ between 10 and 100 Ω^{-1} cm^{-1}. Acrivos finds that if g is taken from the Knight shift, σ is proportional to g^2.

3. The resistivity of the amorphous solid alloy $Ca_{60}Al_{40}$ is 450 Ω cm, probably well above the Ioffe–Regel value, while that of the liquid is 150 Ω cm. Following Howson et al. (1988), quantum interference must be assumed for the former; for the liquid the smaller conductivity leads us to believe that it is absent.

As regards the theoretical discussion, it is assumed that Afonin et al. (1987) are in principle correct, even though in liquids scattering always takes place with some loss of energy, but quantum interference is absent in the Ioffe-Regel regime only because a very large number of jumps are needed for quantum interference. For details of the argument see Kaveh and Mott (1990).

If the considerations given above are correct, we should expect a stronger change of structure at the metal–insulator transition in Cs, where the transition is associated with the critical point, than in mercury. In the former,

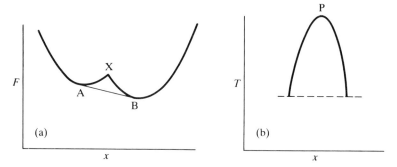

Fig. 8.5. (a) Free energy F at zero temperature as a function of composition x for a liquid undergoing a Mott transition. (b) Solubility gap in metal–ammonia.

also, large fluctuations are expected. Hensel (1990) has described the changes taking place both from fluctuations near the critical point and those resulting from the metal–non-metal transition.

8.4. Comparison between alkali metal vapour and metal–ammonia solutions

Metal-ammonia solutions will not be treated in detail in this book; the author's point of view is given by Mott (1980b, 1991). The one-electron centres are thought to be hydrated electrons. The evidence shows that high concentrations of these centres near the transition form diamagnetic dimers, probably two electrons in one polarization well. The metal–insulator transition might then be thought to be of the band-crossing type (as in mercury), rather than of Mott–Hubbad type. None the less, at the critical point $\sigma \sim \sigma_{min} \sim 100\ \Omega^1\ cm^{-1}$, as for caesium. The model shown in Fig. 4.4 should be appropriate here, the solubility gap occurring as in Fig. 8.5.

9 Vitreous silicon dioxide

9.1. Electronic structure

Vitreous silicon dioxide has a very wide gap of the order of 9 eV and is consequently an excellent insulator. It is important for many reasons; it is the raw material of the communications technology of fibre optics, and in such devices as the MOSFET (metal oxide silicon field effect transistor) layers of vitreous silica are formed on silicon by thermal oxidation (see § 9.3). In this chapter we give a brief account of the band structure of the material, the defects in the network, and the oxidation of silicon by which it can be formed.

In both crystalline silica and the 'perfect' vitreous material each silicon atom has four oxygen neighbours in approximately regular tetrahedral coordination and each oxygen has two approximately equidistant silicon atoms. The Si–O–Si angle is 144° in α-quartz, and has a mean value of 153° in the vitreous material, though it may vary from site to site.† The resulting structure is rather open, allowing diffusion of oxygen (O_2), hydrogen, or sodium ions at sufficiently high temperatures. As in the chalcogenides (Chapter 7) the valence band consists of a comparatively narrow band formed from oxygen lone pair orbitals, and calculations give it a width of ~ 2 eV, separated by more than 2 eV from the much broader band below it, which is formed from the Si–O bonds. They show also that the conduction band has an s-like minimum at $k = 0$.

Measurements of the drift mobility of electrons and holes produced by ionizing radiation have been made. The mobility of an electron at room temperature is high, of order 20 cm² V⁻¹ s⁻¹, and decreases with rising temperature (Hughes 1973, 1977). Any range in energy of localized states must therefore be small compared with $k_B T$. Othmer and Srour (1980) suggested an upper limit of 4.7 meV. Why it should differ in this respect from a-Si is not clear. Mott (1977) suggested that this should follow if the wave function at the bottom of the band is made up primarily of 3s oxygen functions, the situation being then similar to that in liquid rare gases (Chapter 2), but band calculations do not show clearly that this is so showing only that the conduction band minimum is $\mathbf{k} = 0$ and s-like (Schneider and

---

† According to a model of a continuous random network by Bell and Dean (1968), most bond angles lie between 140° and 170°, though tails in their histogram extend to 120° and 180°.

Fowler 1976, 1978; Fowler, 1986). The hole mobility is much smaller, $\sim 10^6 \, \text{cm}^2 \, \text{V}^{-1} \, \text{s}^{-1}$ at 200 K, and at temperatures above this varying as $\exp(-\varepsilon/k_B T)$ with $\varepsilon \simeq 0.16 \, \text{eV}$. This is consistent with a small polaron model (Hughes 1977). Also if a small polaron is formed in a glass, the consequent large effective mass makes it likely that Anderson localization will prevent its motion, leading to an activation energy for motion even at the lowest temperatures. Hughes considered that self-trapping of V_K type was unlikely, the hole remaining on a single oxygen atom.

After about $10^{-7} \, \text{s}$, the activation energy for the motion of a hole was observed by Hughes to *increase* to $\sim 0.35 \, \text{eV}$. This is discussed (Mott 1977) in terms of the delay in forming a polaron of V_K type (Chapter 2). An alternative explanation (Fowler 1983) is that a hole is formed initially in the lower, broader valence band, is weakly trapped there, and takes $10^{-7} \, \text{s}$ to transfer to the upper, narrower part, where it could be self-trapped more strongly. If this is correct, some mechanism is needed by which holes could cross the gap between the two valence bands. For the amorphous material these bands are thought to overlap (Fowler 1986).

Both crystalline and vitreous silicon show a very similar reflectivity spectrum in the ultraviolet, illustrated in Fig. 9.1. The first peak is considered to be a (direct) exciton (Platzöder 1968). On the other hand, calculations show that another band lies below that which gives rise to the direct exciton line at 10.4 eV, but the transition probability to it is vanishingly small, so that it does not show up in Fig. 9.1. The 9.4 eV exciton must dissociate into a free electron in this lower band and a hole.

There is an unsolved problem about the mechanism by which electrons

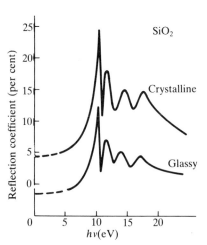

Fig. 9.1. Reflectivity of crystalline and glassy SiO_2 (Phillip 1971).

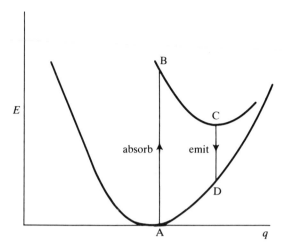

Fig. 9.2. Possible behaviour of exciton energy in SiO_2 as a function of deformation. AB is the energy of the absorption peak, CD of that emitted. q denotes a configurational coordinate, representing movement of atoms.

and holes recombine after excitation. No fluorescence is observed in the neighbourhood of the excitation energy. We have seen that, assuming holes to form polarons, the hopping energy is ~ 0.35 eV and so the binding energy ~ 0.7 eV. On the other hand, luminescence at 2.6 eV is observed, and it has been assumed that, as has also been proposed for chalcogenides (Chapter 7), a very large deformation is possible for the exciton, as in Fig. 9.2 (Mott 1977, 1978a; Trukhin 1980). Self-trapping may be slow, according to the mechanism described in § 6.5 and illustrated in Fig. 9.2, so the exciton can migrate before it is trapped. This deformation implies an enormous Stokes shift (from 9 to 2.6 eV). We see no alternative to the assumption of an attraction between two oxygens, caused by the hole *and* the electron, the electron making all the difference between the weak polaron binding and this large one. The recombination could then be by the mechanism of Dexter *et al.* (1955) (see § 7.7).

9.2. Defects

We turn now to the question of defects in a-SiO_2. The main tool for investigating these is electron spin resonance (e.s.r.) (Griscom 1989). An important e.s.r. signal in SiO_2 is designated E′ (E-prime centre), and this is believed to be the result of a single electron on a silicon dangling bond. In amorphous Si, this could be either a neutral centre at a threefold coordinated silicon, or a positively charged oxygen vacancy; in the crystal only the latter is possible.

It is probable that the defect pair with least energy in SiO_2 is the positively charged threefold coordinated silicon together with a single coordinated negative oxygen ion (the non-bridging oxygen). This, of course, is *not* a valence alternation pair. It has been often proposed that the E-prime centre is produced when an electron is excited and trapped by the former. It now seems more likely, however, that the concentration of those defects is very low, and that the E-prime centre is neutral, and arises from the displacement of an oxygen atom by the ionizing radiation. The mechanism differs from that which produces an e.s.r. signal in the chalcogenides; hard radiation (X-rays or neutrons) are necessary. More than one form of e.s.r. signal is observed according to the nature of the exciting radiation (Griscom 1991) depending perhaps on the distance the oxygen is displaced. A signal from the oxygen is also observed. The signals disappear on annealing.

Other defects are probably the result of impurities; thus a sodium ion leads to an electron trap about 2 eV deep. In soda glasses the charges on the sodium ions are compensated by negatively charged non-bridging oxygens. These shift the absorption edge by about ~ 2 eV towards the red.

That some negatively charged centres exist with levels about 2 eV above the valence band seems to be indicated by the fact that weak absorption, leading to photoconduction, was observed by Appleton *et al.* (1978); it is shown in Fig. 9.3. The absorption here is of the order of 10^{-5} of that in the peak, so centres at this low concentration are indicated.

Griscom (1991) has reviewed the electron spin resonance (e.s.r.) of self-trapped holes; these can be described as forming small polarons, which have high effective mass and so are in Anderson-localized states. Two main types are identified; a hole trapped in the 2p orbital of a normal bridging oxygen, and a hole shared between two valence-band (2p) states in an adjacent oxygen. The distortion here will be different from that in the first case, but essentially here we have a polaron self-trapped by the Anderson mechanism.

9.3. Oxidation of silicon

When silicon, or indeed any other material is oxidized, a film of oxide is formed, through which either the oxidizing agent, or the material to be oxidized, must pass. Films are sometimes crystalline, sometimes amorphous. In silicon oxidized thermally the latter is the case.

Oxide films on silicon are grown either in wet or dry oxygen at temperatures typically 900–1220 °C. In wet oxygen water is the oxidizing agent, oxidation is faster than in dry oxygen, and the oxide contains some

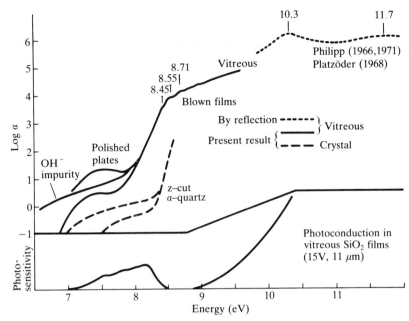

Fig. 9.3. Ultraviolet absorption coefficient α in cm^{-1} in crystalline and vitreous SiO$_2$ at 20 °C, and also photoconduction in vitreous films 11 μm thick for radiation with quantum energy 15 eV (Appleton *et al.* 1978).

water, probably as Si–OH bonds. The rate of oxidation in dry oxygen is proportional to the pressure of oxygen, and in wet oxygen to the partial pressure of water. It thus appears that O$_2$ and H$_2$O diffuse through the oxide layer, and do not dissociate into atoms. In fact for dry oxygen the important work of the group in Paris (Rosenscher *et al.* 1979) showed that, if the heavy isotope of oxygen (^{18}O$_2$) is introduced after some oxide is formed from ^{16}O$_2$, the ^{18}O is deposited at the Si/SiO^2 interface, showing that the O$_2$ molecule passes through the oxide without exchanging an atom with the network already formed. In view of the strength of the Si–O bond this is perhaps not surprising.

As regards the rate of growth for very thin films (say up to 20 Å) which may be transparent to electrons, it is possible that transport of charged oxygen ions is predominant and a mechanism such as that of Cabrera and Mott (1948/49; see also Fehlner and Mott 1970) is operative. Thereafter there are two regimes:

1. The linear regime, where the rate of growth is determined by the reaction of O$_2$ at the Si/SiO$_2$ surface (or of H$_2$O, with liberation of hydrogen). This reaction, unlike that of O$_2$ with a clean surface, has an

activation energy of $\sim 1.5\,$eV. This is somewhat surprising, because the reaction of O_2 with a clean silicon surface is rapid. Doremus (1984) considered the rate of reaction to be slowed down by the stresses. It is suggested that, since the SiO_2 has considerably greater volume per Si atom than the silicon, very large stresses must be set up and the reaction could not proceed if some phase of the (vitreous) oxide were not ductile. Only traces of the stresses remain and are observed.

In this regime, the O_2 must be envisaged as dissolved in the oxide with a concentration proportional to oxygen pressure, so that the rate of growth depends little on thickness.†

2. For greater values of the thickness, denoted by X, parabolic growth is observed. If c is the concentration per unit volume of O_2 at the oxide/oxygen interface, and diffusion is so slow that we suppose that every molecule arriving at the silicon has time to oxidize, then a concentration gradient c/X will exist in the oxide. The rate of oxidation is thus given by

$$\frac{\mathrm{d}X}{\mathrm{d}t} = \frac{c}{X} D\Omega$$

where D is the diffusion coefficient of O_2 in the network and Ω the volume of oxide per O_2 molecule. Integration gives the well-known parabolic law,

$$\tfrac{1}{2}X^2 = (cD\Omega)t.$$

The quantity $cD\Omega$ contains an activation energy of $\sim 26\,$kcal (1.2 eV) and (through c) is proportional to oxygen pressure. For the intermediate case between the linear and parabolic regimes, which covers the formation of films of technical importance, the Deal–Grove equation (Deal and Grove 1965) is appropriate, namely

$$X^2 + AX = B(t + t_0).$$

For H_2O, the activation energy in the parabolic regime is considerably less (0.9 eV). The Paris group (Rigo et al. 1982) showed that in this case, if $H_2{}^{18}O$ is introduced after some film growth with normal water, the ^{18}O is deposited at the oxide/gas interface. It is supposed (Mott 1982) that this is because H_2O can form two Si–OH (silanol) bonds in the SiO_2. These diffuse, and when they recombine to form H_2O, new oxygens, normally ^{16}O, will go into the molecule. It is indeed observed that the solubility of H_2O in SiO_2 is proportional to $p^{1/2}$, when p is the partial pressure of H_2O, so the water must spend most of its time dissociated. But they must diffuse slowly, because the rate of oxidation in the parabolic regime is proportional to p, not $p^{1/2}$.

---

† Mott (1986a) suggested that oxide will grow at kink sites on the silicon surface, being formed therefore in successive monatomic layers. If this is so, stresses resulting from volume change will be small. Stresses may also exist as a result of misfit between a-SiO_2 and Si.

It appears then that it is the H_2O molecules in interstitial sites in the network that move, and their *concentration* is not affected by the fact that they are frequently trapped on the way across the film. In the linear regime, when the growth rate is given by

$$AX = Bt,$$

B/A is proportional to $p^{1/2}$ if O_2 is also present, showing that the Si–OH complex is then active in the oxidation process. A detailed discussion of the oxidation mechanism, particularly of the Paris group, has been given by Mott *et al.* (1989). The model of a continuous random network often used in discussion of vitreous silicon dioxide is criticized in a series of papers by Phillips (1979, 1981, 1985), who considers that the material consists of clusters of crystalline but twinned crystobalite, and that diffusion takes place along the boundaries between these clusters. Such a model gives scope for alteration of the oxide structure during growth, for which there is much evidence. On the other hand, this model is critiziced by Galeenev and Wright (1986).

10 Quasi-amorphous semiconductors

These materials have been extensively studied recently. They have large unit cells, the number of atoms in a unit cell being as large as 1000. Boron and boron-rich compounds are typical. They appear to show hopping conduction at low temperatures, following the Mott law ($\sigma = \sigma \exp\{-(T_0/T)^{1/4}\}$) rather than that expected if the Coulomb gap is important. Figure 10.1 (from Golkova and Tadzkiev 1986; see also Golkova 1991) shows the conductivity and thermopower, with hopping at low temperatures and activated conduction at higher temperatures, for GdB$_{66}$. A full description of their properties is given by Golkova (1989).

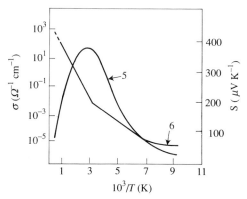

Fig. 10.1. Temperature dependence of thermoelectric power S and conductivity σ (Gulikova 1990).

11 Two-dimensional problems

11.1. Introduction

Scaling theory (§ 3.9) shows that in a two-dimensional system all electronic states are localized. On the other hand much work on conduction in inversion layers (for example Mott *et al.* (1975*b*)) and other two-dimensional systems (for instance delta layers) shows them to behave much like three dimensional metals, with a mobility edge separating localized from extended states, and showing a minimum metallic conductivity $0.1\ e^2/\hbar$. Moreover, the oxide superconductors discussed in Chapter 12 appear to be two-dimensional systems and—if heavily doped—show metallic conductivity at the lowest temperatures and an insulator–superconductor transition which shows all the properties of a mobility edge. We think that this general property of localization in two dimensions has little effect except at the very lowest temperatures, perhaps impossible to reach at present. Moreover, a hypothesis, as yet unproved, is that a mobility edge does exist separating states which are localized exponentially and those with power-law localization. This hypothesis will be discussed in this chapter. We first, however, discuss some phenomena in which metallic behaviour and a mobility are apparently observed.

11.2. Inversion layers

Much of the work on two-dimensional problems has been carried out on currents at the interface between crystalline silicon and the layer of vitreous silicon dioxide formed on it by thermal oxidation. Application of a voltage across the oxide can produce a situation in which the majority carrier at the interface has the sign of the minority carrier in the bulk silicon; the layer where this is so is called an inversion layer; this is illustrated in Fig. 11.1. This is the principle used in the MOSFET (metal oxide silicon field transistor), illustrated in Fig. 11.2; the source–drain current, flowing along the interface, can be measured as a function of the gate voltage across the layer. Schrieffer (1955) first pointed out that under such conditions the energy of the electron perpendicular to the surface must be quantized; the wave

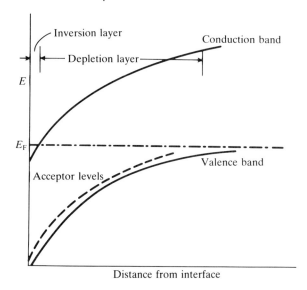

Fig. 11.1. Potential at interface between silicon and silicon dioxide, showing an inversion layer.

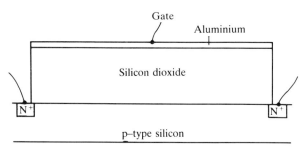

Fig. 11.2. A typical n-channel MOSFET device.

functions ψ could be approximated as

$$\psi = \sin(\pi nx/a) \exp\{i(k_y y + k_z z)\}, \tag{11.1}$$

a being the width of the inversion layer and n an integer, equal to unity in the lowest state. At low temperatures a degenerate electron gas can be formed, all electrons being in states for which $n = 1$, so that perpendicular to the surface (the x-direction) the wave function consists of a single half-wave. Extensive investigations (Fowler *et al.* 1966; Stern 1972; Dorda 1973) confirmed that a two-dimensional gas can form and that higher sub-bands exist in which ψ in the direction perpendicular to the interface consists of two or more half-waves.

The resistivity for a current in the surface layer was early interpreted as being caused by diffuse scattering by surface roughness (Schrieffer 1955) or by charges in the oxide (Fang and Fowler 1968). Either might produce a mobility edge E_c near the bottom of the band. This was assumed to be the case by Mott (1973) and by Stern (1974), who pointed out that a change of gate voltage, by varying the density of electrons in the two-dimensional band, could move the Fermi energy through E_c, leading to a metal–insulator transition of the kind illustrated in Fig. 3.5. The following differences from the three-dimensional case were anticipated:

1. In the hopping regime, instead of the $T^{1/4}$ law (eqn (3.24a)), the conductivity σ was expected to obey the equation

$$\sigma = A \exp(-B/T^{1/3}) \tag{11.2}$$

(Hamilton 1972). This was deduced without the effect of the Coulomb interaction term of Efros and Shklovskii (1975, see § 3.5), which according to them replaces $\frac{1}{3}$ by $\frac{1}{2}$. This is not observed (see Timp et al. 1986).

2. The minimum metallic conductivity should be $0.1\, e^2/\hbar$, a universal constant not depending on any distance.

Experiments by Pepper and co-workers (Pepper et al. 1974a,b,c; for a review see Mott et al. 1975b) confirmed these conclusions, the results being shown in Figs. 11.3 and 11.4. A minimum metallic conductivity appeared to exist, and in the hopping regime the $T^{1/3}$ behaviour was confirmed.

11.3. Localization in two dimensions

Since this work our understanding of two-dimensional systems has been profoundly changed by the scaling theory of Abrahams et al. (1979). These authors pointed out that in two dimensions there is no truly metallic conduction; in the weak disorder limit (see also Gorkov et al. 1980 and Vollhardt and Wölfle 1980a,b), they found for the conductivity σ,

$$\sigma = \sigma_B\{1 - \lambda \ln(L/l)\} \tag{11.3}$$

where σ_B is the Boltzmann conductivity, L and l have the meanings defined in § 3.2, and

$$\lambda = 2/\pi k_F l.$$

If L is the inelastic diffusion length L_i varying as $1/T$, it will be seen that σ drops with decreasing temperature. This behaviour has been widely observed. Bergmann (1983) showed that eqn (11.3) can be obtained as a result of interference between waves scattered by impurities. Since as shown in § 3.9 the β-function plotted against the conductance G is always negative,

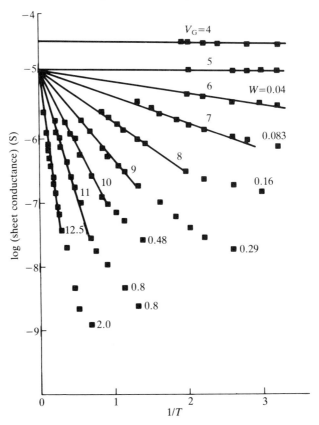

Fig. 11.3. Conductance of n-channel MOSFET device for various values of the gate voltage V_G in volts; W (in meV) is the activation energy $E_c - E_F$.

Abrahams *et al.* deduced that wave functions are always exponentially localized, the localization length being L_0 where

$$L_0 = l \exp(\tfrac{1}{2}\pi l k_F).$$

The conductivity at very low temperatures will then behave like

$$\sigma \sim \exp(-L_i/L_0)$$

when $L > L_0$, in which case eqn (11.3) must break down.

Mott and Kaveh (1981) and Kaveh (1982) made a different proposal, namely:

1. One can write in the quasimetallic regime

$$\sigma = \sigma_B(l/L)^{\lambda};$$

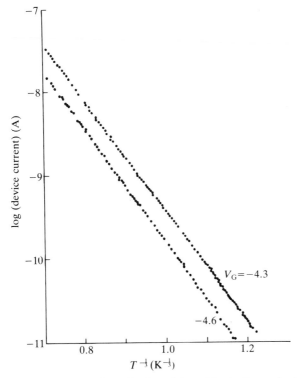

Fig. 11.4. Device current at low T, plotted logarithmically against $1/T$, for two values of the gate voltage in volts, showing $T^{1/3}$ hopping.

eqn (11.3) is then valid for $L \ll L_0$, and thus in all regimes except that of very low temperature. This behaviour follows from the assumption that the states are *not* exponentially localized, but for large r behave as

$$|\psi| \sim 1/r^S;$$

$S < 1$ if when $k_F l \gg 1$, but $S \to \infty$ as $E_F \to E_c$.

2. A mobility edge E_c *does* exist between states with this kind of power-law localization and those with exponential localization, for which the conductivity obeys the hopping formula.

If this model is correct, then no single-valued β-function exists. By deducing L_i from the appropriate formulae, an experimental β-function can be obtained. This was done by Davies *et al.* (1983). Their results are shown in Fig. 11.5 where it will be seen that β certainly is not single-valued. Apart from the results shown in Fig. 11.5, there is, however, no clear experimental

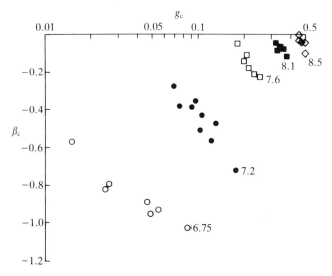

Fig. 11.5. Experimental scaling function β_c as a function of conductance g_c. Quite different curves are obtained for different values of the density of electrons in the inversion layer, shown in 10^{15} m^{-2} for each set of points. g is varied by change of temperature.

evidence to distinguish between the two models.† Except at very low temperatures, in the quasimetallic regime L_i is smaller than the effective radius of the localized state, so that localization hardly affects the conductivity. A mobility edge and minimum metallic conductance $0.1e^2/\hbar$ are expected, except again at very low temperatures.

Extensive numerical work by Haydock and co-workers (Haydock 1986; Godin and Haydock 1988, 1992) has been undertaken to test the power-law assumption, kinks in the density of states are predicted, but the hypothesis is not proved.

11.4. Delta layers

These are layers sufficiently thin to be treated as two dimensional, in the sense of eqn (11.1), produced within an insulator by growth techniques such as molecular beam epitaxy (Koch and Zrenner 1989). We discuss here some work by Mattey *et al.* (1990*a,b*) on boron delta layers in silicon with boron concentration 2×10^{13} to 8×10^{13} cm^{-2}. Using the same concepts as for

---

† Numerical work by Scher and Adler (1985) supports hypothesis 2, while that of Mackinnon and Kramer (1983) supports hypothesis 1.

inversion layers, we expect the behaviour to be similar, with the two-dimensional values of the constants instead of that in doped silicon. Using a magnetic field to contract the orbits, hopping conduction is observed for fields above 12 T of the form $\sigma = A \exp\{-(T_0/T)^{1/2}\}$; unlike the results of Timp *et al.* (1986), a Coulomb gap appears. A metal–insulator transition of the Mott–Hubbard type seems to occur; with weak Anderson localization in the bared tails where the two Hubbard bands overlap (cf. Fig. 4.4). If we use the equation $n^{1/3}a_H = 1/\ln(2zI_0(0))$, z (the coordination number) might be ~ 4 instead of 6 for electrons in three dimensions. For p-type material the absence for n-type materials of the many valleys should not affect the result, and the observed mean distance between centres, at the transition, about half that for three dimensions, has not been explained.

11.5. Quantum Hall effect

The quantum Hall effect, first observed by von Klitzing *et al.* (1980), is perhaps the most striking of the phenomena observed in two-dimensional problems, such as conduction in an inversion layer. If a strong magnetic field H is applied in a direction perpendicular to the *yz*-plane (see eqn (11.1)), the energy spectrum becomes quantized into the Landau levels. For a free electron the Landau levels have energy

$$\varepsilon_n = (n + \tfrac{1}{2})\hbar\omega_c$$

where $\omega_c = He/m$ and $n = 1, 2, \dots \omega_c$ in a periodic lattice will be modified from this formula, and also on account of scattering by disorder, giving a finite lifetime τ, which will broaden the levels by an amount $\hbar/\tau$. A density of states as in Fig. 11.6 is then expected, localized states being shaded and with mobility edges in each band. The resulting Hall current σ_{xy} plotted against gate voltage V_G, which determines n, the density of electrons, will then be as in Fig. 11.7. The flat regions occur when the Fermi energy lies in the localized regime and in these regions σ_{xx} tends to zero. The Hall resistance is then given by nh/e^2, and measurements of the Hall effect have allowed very accurate determination of the fine-structure constant $e^2/\hbar c$. The quantum Hall effect, and particularly the fractional Hall effect, has been the subject of extensive experimental and theoretical work. This is reviewed by Pepper (1985), Prange and Givvin (1987), *Surface Science* (1992), and Aoki *et al.* (1992).

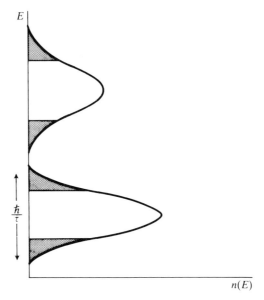

Fig. 11.6. Density of states $n(E)$ resulting from Landau levels; states localized by disorder are shaded.

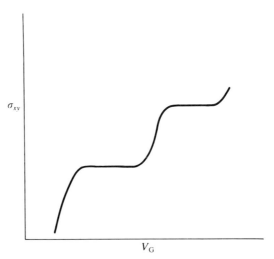

Fig. 11.7. Hall current σ_{xy} as a function of gate voltage V_{G}.

12 High-temperature superconductors

12.1. Introduction

This chapter gives a discussion of the copper oxide superconductors of which $La_{2-x}Sr_xCuO_4$ (LSCO) and $YB_2Cu_3O_{7-\delta}$ (YCBO) are typical. Of these the first is always disordered; LSCO with $x = 0$ is an antiferromagnetic insulator, the magnetic moments being those of copper $3d^9$, of charge-transfer rather than Mott–Hubbard type. When doped with strontium it becomes a p-type semiconductor, with the acceptors in random positions. The carriers are oxygen 2p holes hybridized with Cu 3d, and in common with other p-type semiconductors, should have energies located in an impurity band. Therefore, our understanding of disorder is highly relevant, as it is in materials such as $YB_2(Cu_{1-\alpha}Zn_x)_3O_7$, where there is strong disorder in the copper oxide planes.

The same is true of the n-type materials such as $Nd_{2-x}Ce_xCuO_{4-y}$, in which a carrier is thought to be a $3d^{10}$ state of copper (Toszer *et al.*, 1987) moving through the copper sites for which the configuration is $3d^9$.

Since, as in other superconductors, the current carriers are bosons, in the case of the p-type materials paired holes, to understand many of their properties we need to consider the behaviour of bosons in a non-crystalline environment. In this chapter we concentrate on this aspect of the problem presented by these materials. First, however, we give a brief outline of a theory of their behaviour, which at the time of writing remains controversial.

12.2. A model for copper oxide superconductors

This theory is based on the assumption first made by Shafroth (1955) and in greater detail by Alexandrov and Ranninger (1981*a*,*b*) that superconductors could exist in which the carriers are bosons, and the critical temperature is that at which the boson gas becomes non-degenerate.

In all the copper oxide superconductors, if one plots the transition temperature T_c against the number n of carriers deduced from the chemical composition ($n = x$ in LSCO) T_c rises to a maximum value and then decreases. We denote by n_M this concentration. Figure 12.1 shows schematically this behaviour for LSCO, where n is the concentration of Sr. La_2CuO_4

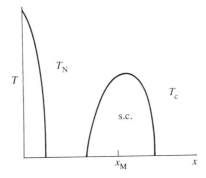

Fig. 12.1. Néel temperature T_N and critical temperature T_c for a copper oxide superconductor as a function of number of positive holes or electrons.

($n = 0$) is an antiferromagnetic insulator with band gap about 2 eV; the moments come from copper ions in the state $3d^9$. The layer structure leads to a Néel temperature T_N—between the planes—of about 200 K, but much higher in the planes (about 1500 K). On doping, the long-range Néel temperature T_N drops to zero as shown also in Fig. 12.1, but inelastic neutron diffraction investigation shows that in the planes some order remains even above T_c. An n increases beyond the point where T_N vanishes, a non-conducting spin glass region appears, in which the spins, if they still exist are anchored in random directions. In this range conduction is, we believe, by variable-range hopping. At a certain value n_T of n an 'Anderson transition' takes place, and the holes become free to move. At that concentration they combine to form pairs, by a mechanism that will be discussed briefly below. We believe this happens to *all* of the carriers, holes or electrons. Apart from any absence of homogeneity, all carriers form bosons at a composition n_T. These give rise to superconductivity. An initial rapid rise in T_c with increasing n is thus expected.

The small correlation length, less than 30 Å, leads us to expect that, at any rate for $n < n_M$ in Fig. 12.1, the pairs form a condensed Bose gas below T_c, but for which, unlike the Cooper pairs of the BCS theory, the overlap between them is small. We assume that the transition temperature T_c is that at which the gas becomes non-degenerate, *not* (as in BCS) that at which the pairs disappear. For $n > n_M$ this may not be so; in this regime much smaller effects of fluctuation are seen. The 'overcrowding effect' of Salje (Chapter 7) may ensure that not all carriers form bosons, so that the electron gas is a mixture of bosons and electrons and the number of bosons decreases with n. We use this concept to discuss the resistivity above T_c in § 12.4.

If this is correct, the copper oxide superconductors must have strong analogies with superfluid ^{4}He. For both, the behaviour bears little relationship to that of a non-interacting gas, for which (in three dimensions)

$$k_B T_c = 3.3 n^{2/3} \hbar^2 / m, \tag{12.1}$$

where m is the mean mass of the boson

$$m = (m_1 m_2 m_3)^{1/3}$$

and n the density of bosons, and the entropy gained at the transition is

$$\Delta S = n k_B \ln 2. \tag{12.2}$$

Prelovsek *et al.* (1987) proposed that the equation should be used for a superconductor, with m_1, m_2, and m_3 the effective masses of a boson in the three crystallographic directions. But for superfluid ^{4}He it is now known that equations of this type (of course with a single mass) are very far from describing the facts. On the other hand, it is known both from theory and experiment that only a small proportion (f) of the helium atoms are in the state $k = 0$; f is about 10 per cent, decreases with volume under pressure, and also with temperature up to T_c (Fig. 12.2). For some references on the helium work, see Mott (1992a) and Alexander and Ranninger (1992).

In helium the entropy up to the transition point, though of the order of $nk_B \ln 2$, is in fact made up of two parts; a part due to phonons, of the type giving specific heat $C_V \sim T^3$, and a much larger part made up by 'rotons' which are now seen as analogous to interstitial atoms moving with a definite k-value, and need an energy to form them which has a lower limit.

For the superconductors the work of Loram *et al.* (1991) on the contribution to the specific heat for which the carriers are responsible gives results as in Fig. 12.3. The entropy of the transition is less by a factor 5 than $nk \ln 2$. This is ascribed by Mott (1992a) to the factor f, together

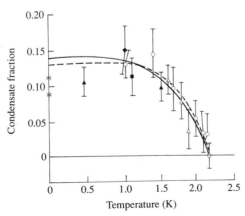

Fig. 12.2. Experimental and theoretical values of condensate fraction in ^{4}He.

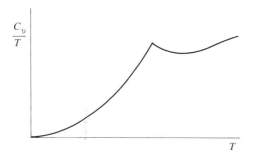

Fig. 12.3. Specific heat of carriers in a high-temperature superconductor (schematic).

with the assumption that roton-like excitations are not formed. De Jongh (1989), however, finds it a result of the two-dimensional structures. For very low temperatures C_V/T may vary as T^n with $n \sim 3$, but an exponential form $\exp(-\Delta S/kT)$ with Δ a constant suggests a gap. However, measurements by Loram *et al.* (1991) of C_v on the system $YBa_2(Cu_{1-x}Zn_x)_3O_7$ reveal the reason for this. Increases in x (zinc concentration) leads to a drop in T_c and in the entropy. In pure $YBCO_{7-\delta}$, with $\delta \simeq 0.6$, the superconductivity disappears and the antiferromagnetic insulator (see Fig. 12.1) forms. At temperatures above T_M the appropriate entropy of the disordered moments is observed. But this disappears when δ decreases and superconductivity appears; obviously the charge carriers responsible for the superconductivity destroy the spin entropy, and a plausible hypothesis is that they do this by forming a spin polaron, with spins oriented antiparallel to the moment on an oxygen 2p hole, as described in § 6.8 (cf. Mott 1992*b*).

This entity carrying a charge of $2e$ must polarize the surroundings. So a dielectric polaron as described in Chapter 6 must be formed. This will affect the effective mass of the polaron.

A mechanism for bonding two such polarons together to form a spinless bipolaron must then be assumed. No clear description of such a mechanism has yet been given, which must involve exchange interaction between copper 3d states as well as the polarization of the medium. We believe that the binding energy is sufficiently large to ensure that there is little dissociation in the normal state up to 500 K or beyond. A detailed discussion by Wood and Cooke (1992) invokes the exchange mechanism as in Fig. 12.4.

Returning now to the effect of adding zinc to YBCO; it is known that the zinc atoms are incorporated in the copper oxide planes. When the zinc content is so large that T_c disappears, C_v/T becomes roughly independent of T, as in Fig. 12.3. The simplest explanation is that the zinc causes Anderson localization of the bosons. Mott (1992) has argued that in this case the bosons behave like fermions, giving

$$C_v = \text{const } N(E_c)k^2T.$$

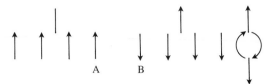

Fig. 12.4. Showing the proposed orientation of spins in a bipolaron. The upper spin on the left is that of the oxygen 2p hole. Outside the polarons, moments are resonating. The antiferromagnetic coupling at AB binds the two polarons together.

The argument is as follows. Because they carry a charge ($2e$) and repel each other, it is certain that the number of bosons in any Anderson *localized* state is limited, perhaps to one only. Therefore, there is an upper limit to the energy E_B in which the states are occupied. We expect then that the properties will be similar to those of a 'Fermi glass' (Chapter 3); C_v/T should be independent of T and conduction by variable-range hopping.

It is possible that the disorder prevents bosons from forming; or gives rise to a mixture of bosons and fermions; if so, as long as states are localized, these conclusions are not altered.

If E_B lies above the mobility edge, E_c, the carriers there can form bosons and, at $T = 0$ will have energies just above E_c. A superconductor will then be present, and T_c will drop to zero with $E_B - E_c$.

12.3. A model for the electrical properties above T_c

Vigren (1973) showed that the diffusion coefficient of a spin polaron should be independent of T (see Chapter 7). The argument is that the entity diffuses a distance a perhaps of order 1 Å each time a moment on the periphery flips from up to down, and that this will occur with frequency ω_N where $\hbar\omega_N \sim k_B T_N$. The diffusion coefficient is thus

$$D \sim \omega_N a^2.$$

This of course does *not* apply to the boson in the condensed state, but only in the non-degenerate gas expected above T_c. For these we can apply Einstein's equation

$$\sigma = nq^2 D/k_B T \qquad q = 2e \qquad (12.3)$$

showing that the resistivity is proportional to T. This is widely observed—though many other theories of this feature have been proposed.

The present author has described the thermopower (Mott 1990a) and the Hall coefficient (Mott 1990b) in terms of this model. For the thermopower

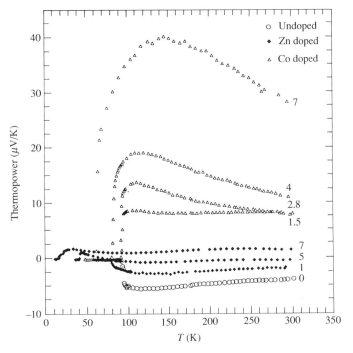

Fig. 12.5. Thermopower of $YB_2(Cu_{1-x}M_x)O_{7-\delta}$ as a function of temperature. M denotes zinc or copper as shown. The numbers against each curve denote $100\times$. (From Obertelli *et al.* 1992.)

S he uses the Heikes (1961) formula

$$S = (k_B/q)\ln\{(1-z)/z\} \qquad (12.4)$$

when $q\,(=2e)$ is the charge on a carrier and z is the ratio of the number of carriers to that of sites available to the bipolarons. In $La_{2-x}Sr_xCuO_4$, as long as the carriers remain in an impurity band, the sites are the Sr atoms; with large Hubbard U, the lower Hubbard band would be fully occupied by polarons and no current is possible. If on the other hand the holes (or electrons) form pairs (bipolarons) $z = \frac{1}{2}$, giving $S = 0$. Actually small ($\ll k_B/e$) values of the thermopower are observed with either sign (Ousef *et al.* 1990). According to Lowe *et al.* (1991), in all untwinned crystals of YBCO it differs in the a and b directions, with values $0.7\,\mu V\,K^{-1}$ and $-6\,\mu V\,K^{-1}$. At higher temperatures it can probably no longer be assumed that the energy of the carriers all lies in an impurity band, so $z < \frac{1}{2}$ and larger values of S occur.

That an equation of type (12.4) is satisfied for fully oxidized YBCO, where the carriers are *not* in an impurity band, suggests that the band for the carriers, which must contain one electron per unit cell with each spin direction, splits into an upper and lower Hubbard band, and without pair formation the lower one must be fully occupied. Figure 12.5 (from Obertelli shows the thermopower of $YBa_2(Cu_{1-x}M_x)O_{7-\delta}$ with M = Zn, Ca, or O (undoped); the figures show $100\times$. It will be seen that here too a constant very small value of the thermopower is found, showing that the carriers are bosons.

A major difficulty in the spin polaron model is that polarized neutron scattering gives no evidence for moments in the superconducting state (Bruckel *et al.* 1987, 1989). Capellmann has suggested that the ferromagnetic spin ordering disappears, leading to a metallic disc in the centre of the polaron. This could, in the author's view, have an effect on the transport properties similar to that of a spin polaron.

References

Abeles, B. and Ping Shen (1974). In *Proceedings 13th Conference on Low Temperature Physics* (ed. K. D. Timmerhaus, W. J. O'Sullivan, and E. F. Hammad), Vol. 3, p. 578. Plenum, New York.

Abeles, B., Ping Shen, Coutts, M. D., and Arie, Y. (1975). *Adv. Phys.* **24**, 407.

Abou-Chacra, R. and Thouless, D. J. (1974). *J. Phys. C: Solid State Phys.* **6**, 1734.

Abrahams, E., Anderson, P. W., Licciardello, D. C., and Ramakrishnan, T. W. (1979). *Phys. Rev. Lett.* **42**, 695A.

Acrivos, J. V. (1972). *Phil. Mag.* **25**, 757.

Acrivos, J. V. and Mott, N. F. (1972). ibid **24**, 19.

Adams, A. R. and Spear, W. E. (1964). *J. Phys. Chem. Solids* **25**, 1113.

Afonin, V. V., Galperin, Yu., Gurevich, V., and Schmitt, A. (1987). *Phys. Rev. A* **36**, 5729.

Aharony, A., Zhang, Y., and Sarachik, M. A. (1992). *Phys. Rev. Lett.* **68**, 3900.

Alexandrov, A. and Ranninger, J. (1981a). *Phys. Rev.* **B 23**, 1706.

Alexandrov, A. and Ranninger, J. (1981b). *Phys. Rev.* **B 24**, 1164.

Alexandrov, A. and Ranninger, J. (1992). *Solid State Commun.* **81**, 403.

Allcock, G. R. (1956). *Adv. Phys.* **5**, 412.

Altshuler, B. L. and Aronov, A. G. (1979). *Solid State Commun.* **30**, 115.

Altshuler, B. L. and Aronov, A. G. (1983). *Solid State Commun.* **46**, 427.

Ambegaokar, A., Halperin, B. I., and Langer, J. S. (1971). *Phys. Rev.* **B4**, 2612.

Anderson, P. W. (1958). *Phys. Rev.* **109**, 1492.

Anderson, P. W., Thouless, D. J., Abrahams, E., and Fisher, C. S. (1980). *Phys. Rev.* **B22**, 3519.

Aoki, H. (1983). *J. Phys.* **C 16**, L2150.

Aoki, H., Tsukada, M., Schluter, M., and Levy, F. (eds) (1992). *New horizons in low-dimensional electron physics*. Kluwer, Amsterdam.

Appleton, A., Chiranjivi, T., and Jafanpour-Chazvini, M. (1978). In *The physics of SiO₂ and its interfaces* (ed. S. T. Pantelides), p. 94. Pergamon, Oxford.

Asbel, Y. (1991). *Phys. Rev.* **B43**, 2435.

Ashcroft, N. and Mermin, N. D. (1976). *Solid state physics*. Holt, Rinehart and Winston, New York.

Baber, W. G. (1937). *Proc. R. Soc.* **A158**, 383.

Baird, B. (1977). *Adv. Phys.* **26**, 657.

Baird, B. (1990). *J. Solid State Chem.* **88**, 28.

Baird, B. (1992). *J. Less Common Metals.* (In press).

Barnes, A. C. and Enderby, J. E. (1985). *J. non-cryst. Solids* **77/78**, 1345.

Bassler, H. (1982). *Phys. Status. Solidi.* **B114**, 561.

Belitz, D. and Schirmacher, W. (1983). *J. non-cryst. Solids* **61/62**, 1073.

Bell, R. J. and Dean, P. (1968). *Phys. Chem. Glasses* **9**, 125.

Benzaquen, M. and Walsh, D. (1984). *Phys. Rev.* **B30**, 7287.

Benzaquen, M., Mazuruk, K., and Walsh, D. (1985). *J. Phys. C: Solid State Phys.* **18**, 1107.

Berggren, K. F. (1974). *J. Chem. Phys.* **60**, 3399.

Bergmann, G. (1983), *Phys. Rev.* **B28**, 2914.

Bergmann, G. (1984). *Phys. Rep.* **107**, (1).

Beyer, W. and Overhof, H. (1984). In *Semiconductors and semimetals*, Vol. 21c (ed. R. K. Willardson and A. C. Beer). Academic Press, New York.

Bhatt, R. N. and Rice, T. M. (1981). *Phys. Rev.* **B 23**, 1920.

Bieri, J. B., Fert, A., Creuzet, G., and Ousset, J. C. (1984). *Solid State Commun.* **49**, 849.

Biskupski, G. (1982). Thesis, Lille.

Biskupski, G., Dubois, H., and Ferré, G. (1981). *Phil. Mag.* **B43**, 183.

Biskupski, G., Wojkiewicz, J. L., Briggs, A., and Remanyi, G. (1984). *J. Phys. C: Solid State Phys.* **17**, L411.

Biskupski, G. and Briggs, H. (1988). ibid. **21**, 883.

Biskupski, G., El Kaanuaski, A. and Briggs, A. (1991). *J. Phys: Condens. Matter* **3**, 8417–24.

Bloch, F. (1982). *Z. Phys.* **52**, 555.

Bogershausen, M. and Micklitz, H. (1992). *Annalen der Physik* **1**, 11.

Bogomolov, V. N., Kudinov, E. K., and Firsov, Yu. A. (1968). *Sov. Phys. Solid State* **9**, 2502.

Bosman, A. G. and Crevecoeur, C. (1966). *Phys. Rev.* **144**, 763.

Brazovskii, S. A. and Kirora, N. N. (1984). Electron self-localization and periodic superstructures in quasi-one-dimensional dielectrics. In *Soviet Scientific Reviews, Section A, Physics Reviews* Vol. 5. Harwood Academic Publishers, Amsterdam.

Brinkman, W. F. and Rice, T. M. (1971). *Phys. Rev.* **B4**, 1566.

Brinkman, W. F. and Rice, T. M. (1973). *Phys. Rev.* **B7**, 1508.

Bruckel, T., Capellmann, H., Just, W., Sharf, O., Kemmler-Sack, S., Kiemel, D., and Schaefer, W. (1987). *Europhys. Lett.* **4**, (10) 1187.

Bruckel, T., Neumann, K. L., Capellmann, H., Sharf, O., Kemmler-Sachs, S., Kiemel, R., and Schaefer, W. (1989). *Solid State Commun.* **70**, 33.

Buckel, W. and Hilsch, R. (1954). *Z. Phys.* **138**, 109.

Buckel, W. and Hilsch, R. (1956). *Z. Phys.* **146**, 27.

Burns, M. J. and Chaikin, P. M. (1985). *J. Phys. C: Solid State Phys.* **18**, L743.

Busch, G. and Labhart, H. (1946). *Helv. Phys. Act* **19**, 463.

Butcher, P. (1976). In *Proc 6th Int. Conf. on Amorphous and Liquid Semiconductors* (ed. B. T. Kolomiets) p. 39, Nauka, Leningrad.

Butcher, P. (1984). *Phil. Mag.* **B50**, L5.

Butcher, P. and Friedman, L. R. (1977). *J. Phys.* **C42**, 3802.

Cabane, B. and Friedel, J. (1971). *J. Physique* **32**, 73.

Cabrera, N. and Mott, N. F. (1948/49). *Rep. Prog. Phys.* **12**, 163.

Cannon, R. D. (1980). *Electron transfer reactions*. Butterworth, London.

Cate, R. C., Wright, J. C., and Cusack, N. E. (1970). *Phys. Lett.* **A 30**, 467.

Chittick, R. C., Alexander, J. H., and Stirling, H. F. (1969). *J. electrochem. Soc.* **116**, 77.

Cochrane, R. W. and Strom-Olsen, J. O. (1984). *Phys. Rev.* **B29**, 1080.

Cohen, M. H., Fritzsche, H., and Ovshinsky, S. R. (1969). *Phys. Rev. Lett.* **22**, 1063.

Cohen, M. H., Economou, E. N., and Soukoulis, C. M. (1983). *J. non-cryst. Solids* **59/60**, 15.

Crevecoeur, C. and de Witt, H. J. (1968). *Solid State Commun.* **6**, 295.

Cutler, M. (1977). *Liquid semconductors*. Academic Press, New York.

Cutler, M. and Mott, N. F. (1969). *Phys. Rev.* **181**, 1336.

Cutler, M. and Rasolondramanitra, H. (1984). *J. non-cryst. Solids* **61/62**, 1097.

Davies, J. H. (1980). *Phil. Mag.* **B41**, 370.

Davies, J. H., Lee, P., and Rice, T. M. (1982). *Phys. Rev. Lett.* **49**, 758.

Davies, R. A., Pepper, M., and Kaveh, M. (1983). *J. Phys. C: Solid State Phys.* **16**, L285.

Davis, E. A. (1984). In *Coherence and energy transfer in glasses* (ed. P. A. Fleury and B. Golding) p. 45, Plenum, New York.

Davis, E. A. and Compton, W. D. (1965). *Phys. Rev.* **A140**, 2183.

Davis, E. A., Michiel, H., and Andraenessens, G. J. (1985). *Phil. Mag.* **B52**, 261.

Deal, B. E. and Grove, A. S. (1965). *J. appl. Phys.* **36**, 3770.

de Gennes, P. G. (1962). *J. Phys. Radium* **23**, 630.

de Jongh, J. (1989). *Physica C* **161**, 631.

Devreese, J. T. (1972). *Polarons in ionic crystals and polar semi-conductors.* North-Holland, Amsterdam.

Dexter, D. L., Klick, C. C., and Russell, G. A. (1955). *Phys. Rev.* **100**, 603.

Dolezalek, F. K. and Spear, W. E. (1970). *J. non-cryst. Solids* **4**, 97.

Dorda, G. (1973). *Festkörperprobleme* **13**, 215.

Doremus, R. H. (1984). *Thin solid films* **122**, 191.

Dougier, P. (1975). Thesis, University of Bordeaux.

Dougier, P. and Casalot, A. (1970). *J. Solid State Chem.* **2**, 396.

Duke, C. and Meyer, R. J. (1981). *Phys. Rev.* **B23**, 2111.

Edwards, J. T. and Thouless, D. (1972). *J. Phys. C: Solid State Phys.* **5**, 807.

Edwards, P. P. and Sienko, M. J. (1978). *Phys. Rev.* **B17**, 2575.

Edwards, P. P. and Sienko, M. J. (1983). *Int. Rev. Phys. Chem.* **3**, 83.

Edwards, S. F. (1961). *Phil. Mag.* **6**, 617.

Edwards, S. F. (1962). *Proc. R. Soc.* **A267**, 528.

Efros, A. L. and Pollak, M. (Eds.) (1985). Electron–electron interactions in disordered systems, Vol. 10 of *Modern problems in condensed matter sciences* (ed. V. M. Agranovich and A. A. Maradudin). North Holland, Amsterdam.

Efros, A. L. and Shklovskii, B. I. (1975). *J. Phys. C: Solid State Phys.* **8**, L49.

El-Hanani, U., Brennert, G. F., and Warren, W. W. (1983). *Phys. Rev. Lett.* **50**, 540.

Elliott, S. R. (1977). *Phil. Mag.* **36**, 1291.

Elliott, S. R. (1980). *J. non-cryst. Solids* **35/36**, 855.

Elliott, S. R. (1984). *Physics of amorphous materials.* Longman Group Ltd., Harlow, U.K.

Elyutin, P. V., Hickey, B., Morgan, G. J., and Weir, G. F. (1984). *Phys. Status Solidi* **B124** (6), 279.

Emin, D. (1973). *Adv. Phys.* **22**, 57.

Emin, D. (1974). *Phys. Rev. Lett.* **32**, 343.

Emin, D. (1975). *Adv. Phys.* **24**, 305.

Emin, D. (1977a). *Phil. Mag.* **35**, 1189.

Emin, D. (1977b). *Solid State Commun.* **22**, 409.

Emin, D. (1984). *Phys. Rev.* **B20**, 5766.

Emin, D., Seager, C. H., and Quinn, R. K. (1972). *Phys. Rev. Lett.* **28**, 813.

Engemann, D. and Fischer, R. (1974). *Proc. 5th Int. Conf. on Amorphous and Liquid Semiconductors* (eds J. Stuke and W. Brenig), p. 947. Taylor and Francis, London.

Engemann, D. (1976). *AIP Conference Proceedings* (ed. G. Lukovsky and F. Galeener) p. 37.

Englman, R. and Jortner, J. (1970). *Mol. Phys.* **18**, 145.

Erydman, A., Cohen, O., and Ovadyahu, Z. (1992). *Solid State Commun.* (in press).

Even, V. and Jortner, J. (1972). *Phil. Mag.* **25**, 705.

Faber, T. E. (1972). *Introduction to the theory of liquid metals.* Cambridge University Press.

Faber, T. E. and Ziman, J. M. (1965). *Phil. Mag.* **11**, 153.

Fang, F. F. and Fowler, A. B. (1968). *Phys. Rev.* **169**, 619.

Fehlner, F. P. and Mott, N. F. (1970). *Oxid. Metals* **2**, 59.

Fenz, P., Muller, H., Overhof, H. and Thomas, P. (1985). *J. Phys. C: Solid State Phys.* **18**, 3191.

Ferré, D., Dubois, H., and Biskupski, G. (1975). *Phys. Status Solidi* **B70**, 81.

Finkelstein, A. M. (1983). *Soviet Physics JETP* **57**, 47.

Fowler, A. B., Fang, F. F., Howard, W. E., and Stiles, P. J. (1966). *Phys. Rev. Lett.* **16**, 901.

Fowler, W. B. (1986). In *Structure and bonding in non-crystalline materials*, ed. G. E. Waldrafen and A. G. Revesz, p. 157. Plenum, New York.

Franz, J. and Davies, J. H. (1986). *Phys. Rev. Lett.* **57**, 475.

Freyland, W., Pfeifer, H. P., and Hensel, F. (1974). *Proc. 5th Conference in Amorphous and Liquid Semiconductors* (ed. J. Stuke and W. Brenig) p. 1327. Taylor and Francis, London.

Friedman, L. and Holstein, T. (1963). *Ann. Phys.* **21**, 494.

Fritsch, G., Willer, J., Wildermuth, A., and Luscher, E. (1982). *J. Phys. F: Metal Phys.* **12**, 2965.

Fritzsche, H. (1960). *Phys. Rev.* **119**, 1899.

Fritzsche, H. (1971). *J. non-cryst. Solids* **6**, 49.

Fritzsche, H. (1978). In *The metal non-metal transition in disordered systems* (ed. L. R. Friedman and D. P. Tunstall), p. 193. Scottish Universities Summer School in Physics.

Fritzsche, H. (1985). *J. non-cryst. Solids* **77/78**, 273.

Fröhlich, H. (1954). *Adv. Phys.* **3**, 325.

Fröhlich, H., Pelzer, H., and Ziemans, S. (1950). *Phil. Mag.* **41**, 221.

Fugol, I. Ya. (1978). *Adv. Phys.* **27**, 1.

Galeener, F. L. and Wright, A. C. (1986). *Solid State Commun.* **57**, 677.

Gehlig, R. and Salje, E. (1983). *Phil. Mag.* **B47**, 229.

Gibbons, D. G. and Spear, W. E. (1966). *J. Phys. Chem. Solids* **27**, 1917.

Glazov, V. M., Chizhevskaya, S. N., and Glagolera, N. N. (1969). *Liquid semiconductors.* Plenum, New York.

Godin, T. J. and Haydock, B. (1988). *Phys. Rev.* **B38**, 5327.

Godin, T. J. and Haydock, B. (1992). *Phys. Rev.* **B46**, 1528.

Golkova, O. A. (1989). *Sov. Phys. Usp.* **32** (8), 665. English translation *Am. Inst. of Phys.* 1990, 665.

Golkova, O. A. (1991). *Chemotronics* **5**, 3.

Golkova, O. A. and Tadzkiev, A. (1986). *J. non-crystalline solids* **87**, 64.

Gorham, E., Bergeron, E. A., and Emin, D. (1977). *Phys. Rev.* **B 15**, 3667.

Gorkov, L. P., Larkin, A. I., and Lee, P. A. (1980). *Phys. Rev.* **B22**, 5142.

Grant, A. J. and Davis, E. A. (1974). *Solid State Commun.* **15**, 563.

Greene, M. P. and Kohn, W. (1965). *Phys. Rev.* **137**, 573.

Greenwood, D. A. (1958). *Proc. Phys. Soc.* **71**, 585.

Grigorovici, R. (1980). *J. non-cryst. Solids* **35/36**, 1167.

Grigorovici, R. (1983). *J. non-cryst. Solids* **59/60**, 221.

Grigorovici, R. and Gartner, P. (1985). *Sir Nevill Mott Festschrift* (ed. D. Adler and H. Fritzsche), Vol. 2 p. 57. Plenum Press, New York.

Griko, J. and Popielowski, J. (1977). *Phys. Rev.* **A16**, 1333.

Griscom, D. C. (1978). In *The Physics of SiO₂ and its interfaces* (ed. S. T. Pantelides) p. 232. Pergamon, New York.

Griscom, D. C. (1989). *Phys. Rev.* **B40**, 4224.

Griscom, D. C. (1991). *Rev. Solid State Science* **4**, 565.

Gubanov, A. I. (1963). *Quantum electron theory of amorphous semiconductors.* Consultants Bureau, New York, 1965.

Gudden, B. and Schottky, W. (1935). *Z. Tech. Phys.* **16**, 323.

Halperin, B. I. and Lax, M. (1966). *Phys. Rev.* **B14**, 722.

Hamilton, E. M. (1972). *Phil. Mag.* **26**, 1043.

Haydock, R. (1986). *Phil. Mag.* **B53**, 554.

Harding, J. H., Masri, P., and Stoneham, A. M. (1980). *J. nucl. Materials* **92**, 73.

Heintze, M. and Spear, W. E. (1986). *J. non-cryst. Solids* **78/79**, 495.

Hensel, F. (1990). *J. Phys. Condensed Matter* **2**, 5433.

Hensel, F. and Franck, E. U. (1968). *Rev. mod. Phys.* **40**, 697.

Hertel, G., Bishop, D. J., Spencer, E. G., Royel, J. M., and Dynes, R. C. (1983). *Phys. Rev. Lett.* **50**, 743.

Herzfeld, K. F. (1927). *Phys. Rev.* **29**, 701.

Hoddeson, L. H. and Baym, G. (1980). *Proc. R. Soc.* **A371**, 8.

Holstein, T. (1959). *Ann. Phys.* **8**, 343.

Howson, M. A. (1984). *J. Phys. F: Metal Phys.* **14**, L25.

Howson, M. A. and Gallager, G. S. (1988). *Phys. Rev.* **170**, 269.

Howson, M. A. and Grieg, D. (1983). *J. Phys. F: Metal Phys.* **13**, L155.

Howson, M. A. and Grieg, D. (1984). *Phys. Rev.* **B30**, 4805.

Howson, M. A., Hickey, B. J., and Morgan, G. J. (1988). *Phys. Rev.* **B36**, 5267.

Hughes, R. C. (1973). *Phys. Rev. Lett.* **30**, 133.

Hughes, R. C. (1977). *Phys. Rev.* **B15**, 2012.

Hung, C. S. and Gleissman, J. R. (1950). *Phys. Rev.* **79**, 726.

Imry, Y. (1980). *Phys. Rev. Lett.* **44**, 469.

Ioffe, A. F. and Regel, A. R. (1960). *Prog. Semiconductors* **4**, 237.

Joanopoulos, J. O. and Lucovsky, G. (Eds) (1984). *The physics of amorphous silicon.* Springer-Verlag, Berlin.

Jones, H., Mott, N. F., and Skinner, H. W. B. (1934). *Phys. Rev.* **45**, 378.

Kamimura, H. and Aoki, H. (1989). *The physics of interacting electrons in disordered systems.* Oxford.

Kastner, M. (1986). *J. non-cryst. Solids* **77/78**, 1173.

Kastner, M., Adler, D., and Fritzsche, H. (1976). *Phys. Rev. Lett.* **37**, 1504.

Kaveh, M. (1982). *J. Phys. C: Solid State Phys.* **30**, L181.

Kaveh, M. (1985). *Phil. Mag.* **B52**, L1.

Kaveh, M. and Mott, N. F. (1981). *J. Phys. C: Solid State Phys.* **4**, L183.

Kaveh, M. and Mott, N. F. (1983). *Phil. Mag.* **B47**, 69.

Kaveh, M. and Mott, N. F. (1984). *Phil. Mag.* **B50**, 175.

Kaveh, M. and Mott, N. F. (1987). *Phil. Mag.* **B55**, 1 and 9.

Kaveh, M. and Mott, N. F. (1992). *Phys. Rev. Lett.* **68**, 1964.

Kawabata, A. (1981). *Solid State Commun.* **38**, 823.

Kawabata, A. (1984). *J. Phys. Soc. Japan* **53**, 1429.

Keem, J. E., Honig, J. M., and Van Zandt, L. L. (1978). *Phil. Mag.* **B7**, 537.

Keesom, W. H. and ʹ.ok, J. A. (1934). *Physica* **1**, 770.

Kimura, T. and Freeman, G. R. (1974). *Can. J. Phys.* **52**, 2220.

Kirkpatrick, S. (1973). *Rev. Mod. Phys.* **45**, 573.

Klement, W., Willencz, R. H., and Duvez, P. (1960). *Nature* **187**, 869.

Klipstein, P. C., Friend, R. H., and Yoffe, A. D. (1985). *Phil. Mag.* **B52**, 611.

Kolomiets, B. T. (1964). *Phys. Status Solid* **7**, 359, 713.

Koon, D. G. and Castner, T. G. (1987). *Solid State Commun.* **54**, 11.

Kosarev, V. V. (1975). *Sov. Phys.-Semicond.* **8**, 897.

Kuusmann, I. L., Liblick, P. K., Liid'ya, G. G., Lushnik, N. E., Lushnik, C. B., and Soovik, T. A. (1976). *Sov. Phys.-Solid State* **7**, 2312.

Lakkis, S., Schlenker, C, Chakraverty, B. K., Buder, R., and Marezio, M. (1976). *Phys. Rev.* **B14**, 1429.

Landau, L. (1933). *Phys. Z. Sov. Un.* **3**, 664.

Landau, L. (1957). *Sov. Phys.-JETP* **3**, 920.

Landau, L. (1959), *Sov. Phys.-HETP* **8**, 70.

Landau, L. and Lifshitz, E. M. (1958). *Statistical physics.* Pergamon, London and New York.

Landau, L. and Pomeranchuk, I. (1936). *Phys. Z. Sov. Un.* **10**, 649.

Laredo, E., Rowan, L. G., and Slifkin, L. (1981). *Phys. Rev. Lett.* **B47**, 384.

Laredo, E., Paul, W. B., Rowan, L., and Slifkin, L. (1983). *Phys. Rev.* **B27**, 2470.

Le Comber, P., Loveland, R. J., and Spear, W. E. (1975). *Phys. Rev.* **B11**, 3124.

Lekner, J. (1967). *Phys. Rev.* **158**, 130.

Lekner, J. (1968). *Phys. Lett.* **A57**, 341.

Linde, J. O. (1931). *Annl. Phys.* **10**, 52.

Linde, J. O. (1932a). *Annl. Phys.* **14**, 353.

Linde, J. O. (1932b). *Annl. Phys.* **15**, 219.

Liu, N. H. and Emin, D. (1984). *Phys. Rev.* **B30**, 3250.

Long, A. P. and Pepper, M. (1984). *J. Phys. C: Solid State Phys.* **17**, 3321.

Lorentz, H. A. (1905). *Proc. Amsterdam Academy* **7**, 684.

Loram, J. W. and Mirza, K. A. (1984). In *Electronic properties of high T_c superconductors.* Springer-Verlag, New York.

Loram, J. W., Mirza, K. A., Liang, W. Y., and Osborne, H. J. (1990). *Physica C* **162–4**, 2430; (1991). *Physica* **162/4**, 493.

Loveland, R. J., Le Comber, P., and Spear, W. E. (1972). *Phys. Rev.* **B6**, 3121.

Lowe, A. J., Regan, S., and Howson, M. A. (1991). *Phys. Rev.* **B44**, 9757.

Ludwig, R. and Micklitz, M. (1984). *Solid State Commun.* **50**, 861.

Ludwig, R., Rasavi, F. G., and Micklitz, H. (1981). *Solid State Commun.* **39**, 363.

Luescher, E., Willer, J., and Fritsch, G. (1983). *J. non-cryst. Solids* **61/62**, 1109.

Lyons, L. E. (1957). *J. Chem. Soc.* 5001.

Mackinnon, A. and Kramer, B. (1983). *Z. Phys.* **B53**, 1.

Maliepaard, H. C., Pepper, M., Newbury, R., and Hill, G. (1988). *Phys. Rev. Lett.* **61**, 367.

Mansfield, R. (1991). In *Hopping transport in solids* (eds M. Pollak and B. I. Shklovskii), p. 469. North-Holland, Amsterdam.

Mattey, N. L., Dowsett, M. G., Parker, E. H. C., Whall, T. E., Taylor, S., and Zhang, J. F. (1990). *Appl. Phys. Lett.* **57** (16), 1648.

Mattey, N. L., Hopkinson, M., Houghton, M., Dowsett, M. G., McPhail, D. S., Whall, T. E., Parker, E. H. C., Booker, G. R., and Whitehurst, J. (1990). *Thin Solid Films* **184**, 15.

Mattey, N. L., Whall, T. E., Biswas, R. G., Kubiah, R. A. A., and Kearney, M. J. (1992). (preprint).

Methfessel, S. and Mattis, D. (1968). *Magnetic Semiconductors in Handbuch der Physik 10/1*. Springer, Berlin.

Micklitz, H. (1985). *Localization and metal-insulator transitions, N.F. Mott Festschrift*, Vol. 3, p. 89, ed. H. Fritzsche and D. Adler. Plenum Press, New York.

Miller, A. and Abrahams, S. (1960). *Phys. Rev.* **120**, 745.

Mochena, M. and Pollak, M. (1991).*Phys. Rev. Lett.* **67**, 109.

Monroe, D. (1985). *Phys. Rev. Lett.* **54**, 146.

Mooij, J. H. (1973). *Phys. Status Solidi* **A17**, 521.

Morrison, S. R. (1980). *Electrochemistry at semiconductor and oxidized metal electrodes.* Plenum Press, New York.

Moss, S. C. and Graczyk, J. F. (1970). *Phys. Rev. Lett.* **23**, 1167.

Mott, N. F. (1936). *Proc. R. Soc. A.* **153**, 699.

Mott, N. F. (1949). *Proc. phys. Soc.* **A62**, 416.

Mott, N. F. (1966). *Phil. Mag.* **13**, 989.

Mott, N. F. (1967). *Adv. Phys.* **16**, 49.

Mott, N. F. (1968). *J. non-cryst. Solids* **1**, 1.

Mott, N. F. (1969). *Phil. Mag.* **19**, 835.

Mott, N. F. (1970). *Phil. Mag.* **22**, 7.

Mott, N. F. (1972). *Phil. Mag.* **26**, 505.

Mott, N. F. (1973). *Electron Power* **19**, 321.

Mott, N. F. (1977). *Adv. Phys.* **26**, 363.

Mott, N. F. (1978*a*). In *Physics of SiO$_2$ and its interfaces* (ed. S. T. Pantelides), p. 1. Pergamon, New York.

Mott, N. F. (1978*b*). *J. non-cryst. Solids* **1**, 1.

Mott, N. F. (1978*c*). *Phil. Mag.* **37**, 377.

Mott, N. F. (1978*d*). *Mat. Res. Bull.* **13**, 1389.

Mott, N. F. (1980*a*). *J. Phys. C: Solid State Phys.* **13**, 5433.

Mott, N. F. (1980*b*). *J. Phys. Chem.* **84**, 1199.

Mott, N. F. (1982). *Phil. Mag.* **A45**, 323.

Mott, N. F. (1984). *Phil. Mag.* **B49**, L75.

Mott, N. F. (1985*a*). *Int. Rev. Phys. Chem.* **4**, 1.

Mott, N. F. (1985*b*). *Phil. Mag.* **B51**, 19.

Mott, N. F. (1985*c*). *Phil. Mag.* **B52**, 169.

Mott, N. F. (1995*d*). *J. non-cryst. Solids* **77/78**, 115.

Mott, N. F. (1986*a*). *S. Afr. J. Phys.* **9**, 41.

Mott, N. F. (1986*b*). *Phil. Mag.* **B53**, 91.

Mott, N. F. (1989). *Phil. Mag.* **B69**, 1985.

Mott, N. F. (1990*a*). *Adv. Phys.* **39**, 55.

Mott, N. F. (1990*b*). *Phil. Mag.* **62**, 273.

Mott, N. F. (1991). *Metal insulator transitions*, 2nd edn. Taylor and Francis, London.

Mott, N. F. (1992*a*). In *High temperature superconductivity*, pp. 271–91, Scottish University Summer School in Physics, Adam Hilger, Bristol.

Mott, N. F. (1992*b*). *Physica C* **196**, 369.

Mott, N. F. and Davies, J. H. (1980). *Phil. Mag.* **B52**, 845.

Mott, N. F. and Davis, E. A. (1079). *Electronic processes in non-crystalline materials*, 2nd edn. Oxford University Press.

Mott, N. F. and Gurney, R. W. (1940). *Electronic processes in ionic crytals*, Ch. 1. Oxford University Press.

Mott, N. F. and Jones, H. (1936). *Theory of the properties of metals and alloys.* Oxford University Press.

Mott, N. F. and Stoneham, A. M. (1977). *J. Phys. C: Solid State Phys.* **10**, 3391.

Mott, N. F. and Kaveh, M. (1981). *J. Phys. C: Solid State Phys.* **14**, L649.

Mott, N. F. and Kaveh, M. (1985a). *Phil. Mag.* **B52**, 177.

Mott, N. F. and Kaveh, M. (1985b). *Adv. Phys.* **34**, 329.

Mott, N. F. and Kaveh, M. (1990). *Phil. Mag. Lett.* **61**, 147.

Mott, N. F., Davis, E. A., and Street, R. A. (1975). *Phil. Mag.* **32**, 961.

Mott, N. F., Pepper, M., Pollitt, S., Wallis, R. H., and Adkins, C. J. (1975). *Proc. R. Soc.* **A345**, 169.

Mott, N. F., Rigo, S., Rochet, F., and Stoneham, A. M. (1989). *Phil. Mag.* **B60**, 189.

Movaghar, B. and Schirmacher, W. (1981). *J. Phys. C: Solid State Phys.* **14**, 859.

Müller, H. and Thomas, P. (1984). *J. Phys. C: Solid State Phys.* **17**, 5337.

Nagels, P., Callaerts, R., and Denayer, M. (1974). *Proc. 5th Conf. on Amorphous and Liquid Semiconductors* (ed. J. Stuke and W. Brenig), p. 867. Taylor and Francis, London.

Nagaev, E. L. (1983). *Physics of magnetic semiconductors.* MIR publishers, Moscow.

Naugle, D. G. (1984). *J. Phys. Chem. Solids* **45**, 367.

Newman, P. F. and Holcomb, D. F. (1983). *Phys. Rev.* **B 18**, 638.

Newsom, D. J. and Pepper, M. (1986). *J. Phys.* **C 19**, 3983.

Nishida, N. J., Furubayarshi, T., Yamaguchi, M., and Ono, K. (1983). *Proc. Tenth Int. Conf. Amorphous and Liquid Semiconductors*, eds K. Tanaki and T. Shimizu, North Holland, Amsteram.

Nordheim, L. (1931). *Annl. Phys.* **9**, 641.

North, D. M., Enderby, J. E., and Egelstaff, P. A. (1968). *J. Phys. C: Solid State Phys.* **1**, 1975.

O'Bryan, H. M. and Skinner, H. W. S. (1934). *Phys. Rev.* **45**, 379.

Obertelli, S. D., Cooper, J. B., and Tullen, J. L. (1992). *Phys. Rev. B.* (In press).

Olin, M. T., Symons, M. C. R., and Eachus, R. S. (1984). *Roc. R. Soc.* **A392**, 227.

Onada, M., Takahashi, T., and Nagasawa, H. (1982). *J. phys. Soc. Japan* **51**, 3868.

Orenstein, T. and Kastner, M. (1981). *Phys. Rev. Lett.* **46**, 1421.

Othmer, S. and Srour, J. R. (1980). *The Physics of MOS insulators* (eds G. Lucovsky) p. 49. Pergamon, Oxford.

Ousef, D. J. and Bryan, H. R. (1990). *Phys. Rev.* **B 41**, 4123.

Ovadyahu, Z. (1986). *J. Phys.* **C 19**, 51, 87.

Overhof, H. (1975). *Phys. Status Solidi* **B67**, 709.

Overhof, H. and Beyer, W. (1980). *J. non-cryst. Solids* **35/36**, 377.

Overhof, H. and Beyer, W. (1981). *Phil. Mag.* **B43**, 1981.

Overhof, H. and Beyer, W. (1983). *Phil. Mag.* **B47**, 377.

Overhof, H. and Thomas, P. (1989). *Electrical transport in hydrogenated amorphous silicon; Springer Tracts in Modern Physics*, Vol. 114.

Paul, W. (1985). *Physics Today* **38** (8), 14.

Paul, W., Lewis, A. J., Connell, G. A. N., and Monstrakas, T. D. (1976). *Solid State Commun.* **20**, 541.

Pauli, W. (1926). *Z. Physik* **41**, 81.

Pepper, M. (1985). *Contemp. Phys.* **26**, 257.

Pepper, M., Pollitt, S., and Adkins, C. J. (1974a). *Phys. Lett.* **A48**, 113.

Pepper, M., Pollitt, S., and Adkins, C. J. (1974b). *J. Phys.: Solid State Phys.* **7**, L273.

Pepper, M., Pollitt, S., Adkins, C. J., and Oakley, R. A. (1974c). *Phys. Lett.* **A47**, 71.

Perron, J. C. (1967). *Adv. Phys.* **16**, 657.

Phillip, H. R. (1966). *Solid State Commun.* **4**, 73.

Phillip, H. R. (1971). *J. Phys. Chem. Solids* **32**, 1935.

Phillips, J. C. (1979). *J. non-cryst. Solids* **34**, 153.

Phillips, J. C. (1981). *Phys. Rev.* **B24**, 1744.

Phillips, J. C. (1983). *Solid State Commun.* **47**, 191.

Phillips, J. C. (1985). *Phys. Rev.* **B32**, 5356.

Phillips, J. C. (1990). *Phys. Rev. Lett.* **64**, 107.

Ping Shen and Klafter, J. (1983). *Phys. Rev.* **B27**, 2583.

Platzöder, K. (1968). *Phys. Status Solidi* **29**, K. 63.

Pollak, M. (1972). *J. non-cryst. Solids* **11**, 1.

Pollak, M. (1992). *Phil. Mag.* **B 65**, 657.

Pollak, M. and Adkins, C. J. ibid 855.

Pollert, E., Hejtmaneck, J., Doumerc, J. P., Clavene, J., and Hagenmuller, P. (1983). *J. Phys. Chem. Solid* **44**, 273.

Pope, M. and Swenberg, C. E. (1982). *Electronic processes in organic crystals.* Oxford University Press.

Powell, A. R., Mattey, N. L., Kubiak, R. A. A., Parker, E. H. C., Whall, T. E., and Bowen, D. K. (1991). *Semicond. Sc. Technol.* **8**, 227.

Prelovsek, P., Rice, T. M., and Zhang, F. C. (1987). *J. Phys.* **C 20**, L289.

Rashba, E. I. (1982). In *Excitons* (ed. E. I. Rashba and M. D. Sturge), Chapter 13. North-Holland, Amsterdam.

Rashba, E. I. (1985). In *Excitons at high densities* (ed. H. Haken and S. Nikitine), p. 150. Springer Tracts in Modern Physics, Vol. 73, Springer-Verlag, Berlin.

Rice, T. M. (1977). *Solid State Phys.* **32**, 1.

Rigo, S., Rochet, B., Agius, B., and Straboni, A. (1982). *J. Electrochem. Soc.* **129**, 867.

Rosenbaum, T. F., Andres, K., Thomas, G. A., and Bhatt, R. N. (1980). *Phys. Rev. Lett.* **43**, 1723.

Rosenscher, E., Straboni, A., Rigo, S., and Amsel, G. (1979). *Appl. Phys. Lett.* **34**, 259.

Salje, E. H. K. (1990). *Phil. Mag. Lett.* **62**, 277.

Salje, E. H. K. and Guttler, B. (1984). *Phil. Mag.* **B56**, 607.

Sawatzky, G. A. and Allen, J. W. (1984). *Phys. Rev. Lett.* **53**, 2339.

Sayer, M., Chen, R., Fletcher, R., and Mansingh, A. (1975). *J. Phys. C: Solid State Phys.* **8**, 2059.

Schein, L. B. and McGhie, A. R. (1979). *Phys. Rev.* **B20**, 1631.

Scher, G. M. and Adler, D. (1985). In *Localization and metal-insulator transitions* (ed. H. Fritzsche and D. Adler), p. 441. Plenum Press, New York.

Scher, H. and Montroll, E. W. (1975). *Phys. Rev.* **B12**, 2455.

Schiff, L. I. (1955). *Quantum mechanics*, 2nd edn. McGraw-Hill, New York.

Schirmacher, W. (1991). *Phys. Rev.* **B41**, 2461.

Schirmer, O. F. and Scheffler, M. (1982). *J. Phys. C: Solid State Phys.* **15**, L645.

Schlenker, C. and Marezio, M. (1980). *Phil. Mag.* **B42**, 453.

Schneider, P. M. and Fowler, W. B. (1976). *Phys. Rev. Lett.* **30**, 625.

Schneider, P. M. and Fowler, W. B. (1978). *Phys. Rev.* **B17**, 1302.

Schonherr, G., Schmutzler, R. W., and Hensel, F. (1979). *Phil. Mag.* **B40**, 411.

Schreiber, M., Kramer, B., and Mackinnen, A. (1989). *Physics Scripta* **T89**, 67.

Shafarman, W. N. and Castner, T. G. (1986). *Phys. Rev.* **B33**, 3570.

Shafarman, W. N., Castner, T. G., Brooks, J. S., Marks, K. P., and Naughton, J. P. (1986). *Phys. Rev. Lett.* **56**, 980.

Shafarman, W. N., Koon, D. G., and Castner, T. G. (1989). *Phys. Rev.* **B40**, 1216.
Shapiro, B. (1984). *Phil. Mag.* **B 50**, 241.
Shapiro, B. and Abrahams, E. (1981). *Phys. Rev.* **B 24**, 4025.
Shklovskii, B. I. and Efros, A. L. (1984). *Electronic properties of doped semiconductors.* Springer-Verlag, Berlin.
Shlimak, I. S. (1990). *Hopping and related phenomena*, eds. H. Fritzsche and M. Pollak, p. 49, World Scientific Singapore.
Shockley, W. (1950). *Electrons and holes in semiconductors.* Van Nostrand, Princeton, New Jersey.
Silinsh, E. A. (1980). *Organic and molecular crystals.* Springer-Verlag, Berlin.
Silinsh, E. A. and Jurgis, A. J. (1985). *Chem. Phys.* **94**, 77.
Siran, V., Entin-Wohlmann, O., and Imry, Y. (1988). *Phys. Rev. Lett.* **50**, 1566.
Song, K. S. (1969). *J. Phys. Soc. Japan* **26**, 1131.
Spear, W. E. (1974). *Adv. Phys.* **23**, 523.
Spear, W. E. (1983). *J. non-cryst. Solids* **59/60**, 1.
Spear, W. E. and Le Comber, P. G. (1972). *J. non-cryst. Solids* **8/10**, 727.
Spear, W. E. and Le Comber, P. G. (1975). *Solid State Commun.* **17**, 1193.
Spear, W. E. and Le Comber, P. G. (1982). *Topics appl. Phys.* **55**, 63.
Srivastava, V. (1989*a*). *J. Phys. Cond. Matter* **1**, 439.
Srivastava, V. (1989*b*). *Phys. Rev.* **B 4**, 5667.
Srivastava, V. and Weaire, D. (1978). *Phys. Rev.* **B18**, 6635.
Stern, F. (1972). *Phys. Rev.* **B5**, 4891.
Stern, F. (1974). *Phys. Rev.* **B9**, 2762.
Street, R. A. (1976). *Adv. Phys.* **25**, 379.
Street, R. A. (1980). *Phys. Rev.* **B21**, 5775.
Street, R. A. (1982). *Phys. Rev. Lett.* **69**, 1187.
Street, R. A. (1991). *Hydrogenated amorphous silicon.* Cambridge University Press.
Street, R. A. and Biegelson, D. K. (1980). *J. non-cryst. Solids* **35/36**, 357.
Street, R. A. and Mott, N. F. (1975). *Phys. Rev. Lett.* **35**, 1293.
Stutzmann, M. and Street, R. A. (1985). *Phys. Rev. Lett.* **54**, 186.
Stutzmann, M. and Stuke, J. (1983). *Solid State Commun.* **43**, 635.
Sumi, H. (1972). *J. phys. Soc. Japan* **33**, 327.
Sumi, H. (1978). *Solid State Commun.* **28**, 309.
Sumi, H. (1979*a*). *J. Chem. Phys.* **70**, 3775.
Sumi, H. (1979*b*). *J. Chem. Phys.* **71**, 3463.
Summerfield, S. and Butcher, P. N. (1985). *J. non-cryst. Solids* **77/78**, 135.
Theye, M. L., Gheorghiu, A., and Rappeneau, T. (1980). *J. Physique* **41**, 1173.
Theye, M. L., Gheorghiu, A., Driss-Kkodja, K., and Boccara, C. (1985). *J. non-cryst. Solids* **77/78**, 1293.
Tiedje, T. and Rose, A. (1980). *Solid State Commun.* **37**, 48.
Tiedje, T., Moustakas, T. D., Morel, D. L., Cebulka, J. M., and Abeles, B. (1981). *J. Phys. Paris* **4**, 155.
Tiedje, T. and Rose, A. (1980). *Solid State Commun.* **37**, 48.
Thomas, G. A. (1983). *Physica* **B117**, 81.
Thomas, P. (1985). *J. non-cryst. Solids* **77/78**, 121.
Thouless, D. J. (1977). *Phys. Rev. Lett.* **39**, 1167.
Timp, G., Fowler, A. B., Hartstein, A., and Butcher, P. N. (1986). *Phys. Rev.* **B53**, 1499.
Toszer, S. W., Kleinsusser, A.W., Penney, T., Kaiser, D., and Holtzberg, F. (1987). *Phys. Rev. Lett.* **59**, 1768.

Triska, A., Dennison, D., and Fritzsche, H. (1975). *Bull. Am. Phys. Soc.* **20**, 392.

Trukhin, A. N. (1980). *Phys Status Solidi* **B98**, 541.

Tsang, C. and Street, R. A. (1979). *Phys. Rev.* **B19**, 3027.

Tsuda, N., Nasi, K., Yanaso, A., and Siratori, K. (1990). In *Electronic conduction in oxides*, Springer Series in Solid State no 94.

Turkevich, L. A. and Cohen, M. H. (1984). *Phys. Rev. Lett*, **53**, 2323.

Vardeny, Z. and Tauc, J. (1985). *Phys. Rev. Lett.* **54**, 1044.

Van Elp, J. (1990). *Electronic structure of doped transition metal alloys.* Thesis, Groningen.

Vigren, P. T. (1973). *J. Phys.* **C 2**, 967.

Vilfan, J. (1973). *Phys. Status Solidi* **B59**, 357.

Vollhardt, D. and Wölfle, P. (1980*a*). *Phys. Rev. Lett.* **45**, 842.

Vollhardt, D. and Wölfle, P. (1980*b*). *Phys. Rev.* **B22**, 4666.

Von Klitzing, K., Dorda, G., and Pepper, M. (1980). *Phys. Rev. Lett.* **45**, 494.

Warren, W. W. (1970*a*). *J. non-cryst. solids* **4**, 168.

Warren, W. W. (1970*b*). *Solid State Commun.* **8**, 1269.

Warren, W. W. (1972*a*). *Phys. Rev.* **B3**, 3708.

Warren, W. W. (1072*b*). *J. non-cryst. Solids* **8–10**, 211.

Weaire, D. and Srivastava, V. (1977). *Amorphous and liquid semiconductors*, ed. W. E. Spear, Scottish Summer School in Physics, Edinburgh, p. 286.

Whall, T. E., Yeung, K. K., Progkova, Y., and Braben, V. A. N. (1984). *Phil. Mag.* **B50**, 689; (1986) ibid **B54**, 505; (1987) ibid **B56**, 99.

Willer, J., Fritzsch, G., Rausch, W., and Luscher, E. (1983). *Z. Phys. B* (Condensed Matter) **50**, 39.

Wilson, A. H. (1931). *Proc. R. Soc.* **A133**, 458; **A134**, 277.

Wood, R. F. and Cooke, J. F. (1992). *Phys. Rev.* **B 45**, 5585.

Yakovlev, D. R., Uraltsev, I. N., Ossau, W., Bicknell, T., Tassung, R. N., Waag, A., and Schmensser, S. (1992). *Solid State Commun.* **82**, 29.

Yamaguchi, M., Morigaki, K., and Khimoto, H. (1985). *Solid State Commun.* **28**, 81.

Yonezawa, E. (ed.) (1980). *Fundamental physics of amorphous semiconductors.* Springer-Verlag, Berlin.

Zallen, R. (1983). *Physics of amorphous solids.* John Wiley and Sons, New York.

Zhang, F. C. and Rice, T. M. (1988). *Phys. Rev.* **B37**, 3759.

Ziman, J. M. (1961). *Phil. Mag.* **6**, 1013.

Zunt, T., Rohda, M., and Micklitz, H. (1990). *Phys. Rev.* **B21**, 4831.

Zvyagin, I. P. (1973). *Phys. Status Solidi* **B58**, 443.

Subject index

amorphous metals 14ff, 62
Anderson localization 19ff, 49, 61

bipolarons 77
bipolarons in superconductors 127ff

chalcogenide glasses 6, 34, 98ff
copper–zinc alloys 15
copper-oxide superconductors 127ff
correlation length 128
Coulomb gap 33, 42

dangling bonds 86ff
drift mobility 96
doping efficiency in amorphous silicon 90

eight minus N rule 83
electron–electron interaction, effect of 56
electron spin resonance 115

Fermi–Dirac statistics 1
Fermi energy 25
Fermi energy – pinning of 99
Fermi glass 35
Fermi surface 2, 4

gado lineum sulphide 82
gap in energy spectrum of glasses 85
glasses, metallic 14
granular metals 35

Hall effect in amorphous silicon 75
Hall effect in liquid metals 12
Hall effect in polarons 70
Hall effect in superconductors 131
helium, liquid 13
helium, liquid analogy with
  superconductors 129
hopping conduction 31ff, 38
Hubbard U 47
Hubbard U bands 57

impurity conduction 5, 52
indium phosphide 42
inelastic diffusion length 23
inversion layers 119
Ioffe–Regel condition 5, 21

Knight shift 24, 107, 108
Kubo–Greenwood formula 17

Landau–Baber scattering 23
lead, liquid 11
liquid metals 8, 104ff
liquid rare gases 12
localization 19, 25
localization length 27
localization in two dimensions 121

magnesium–zinc alloys 15
magnetoresistance 23, 40
mercury, liquid 103ff
metal-ammonia 110
metal–rare gas systems 37, 61
Meyer–Nelden rule 93
minimum metallic conductivity 26, 40
mobility edge 19, 25, 26
Mott–Hubbard transition 38, 49, 54
Mott–Hubbard transition in liquids 104
multiple scattering 22

nickel oxide 73

Ornstein–Zernike formula 11
oxidation of silicon 114
oxide superconductors 128

pair distribution function 9
palladium–zirconium alloys 17
paramagnetism in liquid metals 107
percolation transitions 37
p–n junctions 99
polarons 66ff
polarons, large 71

Author index